Naturalists' Handbooks 29

Aphids on deciduous trees

TONY DIXON
School of Biological Sciences,
University of East Anglia,
Norwich NR4 7TJ,
UK

and

THOMAS THIEME
Birkenallee 19,
18184 Sagerheide,
Germany

Published for the Company of Biologists Ltd by

The Richmond Publishing Co. Ltd

P.O. Box 963, Slough SL2 3RS

Series editors
S. A. Corbet and R. H. L. Disney

Published by The Richmond Publishing Co. Ltd,
P.O. Box 963, Slough, SL2 3RS
Telephone: 01753 643104
email: rpc@richmond.co.uk

Text © The Company of Biologists Ltd 2007

ISBN 978 0 85546 314 4 Paper

Contents

Editors' preface

Aphids are tempting subjects for investigation because they are often abundant, easily found, and unlikely to run away, and because of intriguing facets of their biology such as polyphenism, parthenogenesis, gall formation and the production of a soldier morph. However, until now identification has presented a severe challenge to the beginner, because there are so many rather similar species, many of which have several different forms. We have long wanted a Naturalists' Handbook that would make the study of aphids more accessible to those unfamiliar with the group. We are delighted that Tony Dixon and Thomas Thieme have met this challenge.

By confining themselves to aphids living on broad-leaved trees, they have reduced problems of identification to manageable proportions, and we hope this book will encourage more field studies of this important and ubiquitous group of insects. The coloured illustrations in the other books in this series are all paintings. At the authors' suggestion, we have departed from that tradition for this title, and we feel that these excellent colour photographs, showing both the aphids and their effects on the host plants, will be a superb aid to identification.

S.A.C.
R.H.L.D.

Acknowledgements

It is impossible to create a key for the identification of aphids without being influenced by other keys. Those presented are based on keys published by Blackman & Eastop (1994); Heie (1980, 1982, 1986, 1991, 1993, 1995); Shaposhnikov (1964); Stroyan 1977, 1984); and Thieme & Müller (2000). The drawings for Fig. 24b–d are modifications of drawings published by Vidano (1959).

The authors gratefully thank Professor S. Barbagallo (University of Catania), Dr. U. Heimbach (Federal Biological Research Centre for Agriculture and Forestry, Braunschweig), Dr. N.P. Hidalgo (University of Leon), Dr. G. Hopkins (Hopkins Inverterbate Surveys, Norwich), Dr. A. Pollini (Plant Disease Laboratory, Bologna) and K. Schrameyer (Amt für Landwirtschaft, Landschafts- und Bodenkultur Heilbronn) for providing the photographs listed below (the first number indicates the plate, the second the picture):
Prof. Barbagallo: 5.4; 7.3; 9.5 and 6; 16.2 and 3
Dr. Heimbach: 15.1 and 2
Dr. Hidalgo: 5.2 and 3; 10.1 and 2; 15.3, 4, 5 and 6
Dr. Hopkins: 3.1
Dr. Pollini: 6.5; 14.1
Mr. Schrameyer: 1.1 and 3; 2.5 and 6; 3.2 and 6; 4.1 and 5; 5.6; 6.2, 3 and 4; 8.1, 5 and 6; 9.1, 2 and 3; 10.4, 5 and 6; 11.1, 2 and 4; 12.1 and 4; 13.1, 2, 3, 4, 5 and 6; 14.3; 16.1, 4, 5 and 6. We also thank Simon Leather for permission to reproduce Figs. 33 and 34.

We are especially indebted to K. Schrameyer who was very helpful in finding interesting aphids in the field and who in addition to his beautiful photographs contributed much information on their natural history. It is a pleasure to acknowledge our indebtedness to Sally Corbet and Henry Disney for their encouragement, and in particular to Sally for her meticulous editing.

A.F.G.D.
T.T.

1 Introduction

Aphids are herbivorous insects that feed on the phloem sap of plants. They do this by inserting the fine threadlike mouthparts, the stylets, deep into the tissues of plants until they locate and pierce the phloem elements, which transport dissolved nutrients throughout plants. Phloem sap is under very high pressure, which forces it up the fine channel in the stylets and into the aphid's stomach. By feeding in this way aphids generally cause very little damage to the foliage of plants and therefore even very large populations often go unnoticed.

The more obvious distinguishing features of aphids are illustrated in fig.1. (A) The stylets are sheathed by a beak-like rostrum, the base of which lies between and behind the bases of the fore legs; (B) each antenna consists of two short thick basal segments and a thinner flagellum, of at most four segments, the last of which consists of a thicker basal part and a thinner part nearer the tip; (C) there are three transparent facets situated behind each compound eye; (D) each foot is made up of two segments; (E) the wings have only one prominent longitudinal vein; and (F) there is a pair of tubes (siphunculi) on the fifth abdominal segment.

Why is this account restricted to aphids living on deciduous trees? Most of these aphids are host specific and many belong to the subfamily Drepanosiphinae, species of which are first recorded as fossils in the late Cretaceous, more than 70 million years ago. So the Drepanosiphinae is a relatively old group and as a consequence the extant species are taxonomically well defined. In addition, the species in this subfamily are morphologically less variable than those in the more recently evolved Aphidinae and Lachninae, which make up most of the present day aphid fauna. Deciduous trees are also familiar and therefore easily identified. From an ecological and practical point of view this is important as it means that these aphids and their hosts are easily found and identified, even in the field.

Why study aphids rather than other insect herbivores? After a little experience aphids can be reared and handled with ease, and as they are common their behaviour and ecology can be studied in the field. Above all, however, their fast pace of life means that within the short span of time available to students for practical work aphids can complete several generations and become very abundant. This has great advantages for anyone designing experiments to test behavioural or ecological concepts.

Fig. 1. Diagnostic morphological features of aphids (see text).

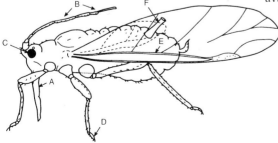

2 Aphids

There are nearly 5,000 species of aphids in the world, most of which live in the temperate regions. Although some species feed on the roots, stems and even the trunks and branches of trees, most feed on the leaves or needles of plants. They have colonized a wide range of terrestrial habitats both above and below ground, and even aquatic habitats, where many live on the submerged parts of freshwater plants and some live on the lower parts of the stems of salt marsh plants or on mosses, and are temporarily inundated when the tide comes in or when it rains. These species breathe air through pores (spiracles) on each side of the body, just like all other aphids, but they extract oxygen from a bubble of air held by water repellent hairs, spines and tubercles that form a cuticular meshwork over certain parts of their bodies. Water repellent hairs around the spiracles ensure that the bubble of air has continuity with the air in the respiratory system of the aphid. The volume of the bubble is fixed because the cuticular meshwork is stiff and does not collapse. Since the surface area of the bubble is large, gas exchange between it and the surrounding water is rapid. When oxygen is used by the aphid a small drop in gas pressure occurs within the bubble and more oxygen enters from the surrounding water.

Distribution

Although most aphid species live in temperate regions, some are found in the Arctic, the Antarctic and the tropics. The subfamilies Aphidinae and Drepanosiphinae, which make up 70% of modern aphid species, are not restricted to a particular region. Only the subfamilies Greenideinae and Hormaphidinae, which make up 7%, are so restricted, and most of the species are currently found only in South East Asia and Australia. However, the presence of fossil Greenideinae in the Balkans indicates that this group once had a much wider distribution. As the most important subfamilies of aphids are not restricted in their distribution we may ask why there are so few species of aphid in the tropics compared to the temperate regions. This is in marked contrast to most taxa, which show the reverse latitudinal gradient in species diversity with most species occurring in the tropics. This topic is discussed in more detail on page 42.

Evolution

Although the oldest fossil aphid, the wing of *Triassoaphis culitus*, is from the Triassic, 190 million years ago, it is likely that the superfamily to which aphids belong (Aphidoidea) evolved in the Carboniferous, 280 million years ago, from insects that lived on primitive conifers

left hind leg   rostrum

antennae

Fig. 2. Immature wingless individual of *Germaraphis* embedded in Baltic amber.

(gymnosperms). The major diversifications in aphids and flowering plants (angiosperms) appear to have occurred simultaneously in the early Cretaceous some 120 million years ago. This, and the fact that most extant aphids live on angiosperms, indicates that the diversification of aphids is closely linked with that group of flowering plants.

The most convincing fossil aphids are those trapped in Baltic amber (fig. 2), the petrified resin of coniferous trees. The many insects, including aphids, entrapped and preserved in this material, like the humans petrified at Pompeii, are very lifelike in their posture, and some are remarkably like modern aphids. However, the initially free flowing and very sticky resin that entrapped the aphids was produced 50 to 35 million years ago.

Size

Aphids vary in length from 0.7 to 7mm. Those that feed on the deeply located phloem elements in the trunks of trees are larger than those that feed on the more accessible phloem elements of leaves. To reach phloem elements deep within a trunk of a tree an aphid needs long stylets. They are not coiled up within the body, as in scale insects, but are characteristically enclosed within a rostrum and exposed by withdrawing the tip of the rostrum into its basal segments. The largest aphid, *Stomaphis quercus*, which feeds on the trunks of mature oak trees, has a rostrum that is nearly twice as long as its body. When the rostrum is withdrawn to expose the stylets, the invaginated rostrum runs internally the full length of the body. Thus the structure of the rostrum limits the length of the stylets, making it physically impossible for small aphids to feed on tissues very deep within a plant.

The range in size, with the large species of aphids feeding on the trunks of trees and small species feeding on leaves, is well illustrated by the aphids feeding on oak (fig. 3). The positive relationship between rostrum length and

Fig. 3. Feeding positions of five species of aphids living on oak (a), and the relationship between the length of the stylets and the body plotted on a logarithmic scale (b). (1) *Tuberculatus annulatus*; (2) *Thelaxes dryophila*; (3) *Lachnus roboris*; (4) *Lachnus iliciphilus*; (5) *Stomaphis quercus*.

(a)

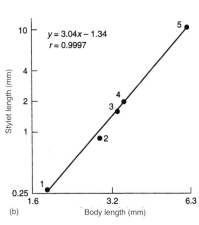

$y = 3.04x - 1.34$
$r = 0.9997$

Stylet length (mm)

Body length (mm)

(b)

body length shown by aphids living on oak also holds for
aphids in general. Within a species, successive
developmental stages (instars) feed on progressively larger,
deeper veins. This is associated with an increase in rostrum
length with age, and the first instar has a longer rostrum
relative to its body length than an adult. Thus both within
and between species the size of an aphid is correlated with
the depth of the phloem elements on which it feeds.
Evidently the depth of the phloem constrains the minimum
birth size. If there is a directly proportional linear
relationship between birth size and adult size, a doubling of
body length at birth will result in a doubling of adult body
length. We find that the optimum adult body length is on
average 16 times the body length at birth.

Food

Aphids feed on the contents of living cells, phloem
elements. When an element is punctured by an aphid's
stylets there is a drop in pressure within it, and one would
expect the plant to respond as it does when wounded and
seal off the element. However, aphids inject saliva into plants
as they penetrate tissues on the way to the phloem and also
salivate into the phloem elements. Experiments using
radioactive aphids indicate that substances in their saliva are
rapidly translocated throughout a plant from the topmost
leaves to the smallest roots. The saliva probably acts like the
anti-coagulant used by blood-sucking insects; perhaps aphid
saliva prevents the plant sealing off the elements that have
been punctured by aphids. Feeding in this way, most aphids
appear to do very little damage to plants, but the saliva of a
few species can dramatically affect the growth of plants,
inducing them to produce galls. Simple galls or distorted
leaves (pseudo-galls) result when components in aphid
saliva inhibit the elongation of the stem or cause the leaves to
twist. The resultant bunch of twisted leaves provides the
aphid with shelter and a food source that remains favourable
for longer than ungalled leaves. In the case of true galls the
initiating aphid is an architect in that it carefully chooses
where and when to inject saliva into a plant and so induces it
to produce a complex and elegant structure that completely
encloses the aphid and its descendants. The aphid even
controls the subsequent opening of the gall so that its
descendants can escape and move elsewhere.

The food of aphids, phloem sap, is rich in sugars but
poor in amino nitrogen essential for growth. This poses
problems for aphids. First, the osmotic concentration of
phloem sap is greater than that of aphid body fluids. This
should result in water moving along an osmotic gradient
from the body into the gut of aphids, resulting in their
desiccation. Aphids surmount this problem by quickly
converting the mainly simple sugars in the incoming food
into complex sugars. This reduces the number of molecules
in solution and the osmotic concentration of the food.

Secondly, the very low concentration of amino nitrogen in their food means they have to process large volumes of sap if they are to obtain sufficient amino nitrogen to sustain their rapid growth. Adult sycamore aphids process at least their own weight of phloem sap per day, and immature aphids process several times their own weight. They can assimilate the nitrogen from phloem sap with an efficiency of over 60%, despite the need to process large volumes of food and the absence of a mechanism for concentrating dietary nitrogen before assimilation. This is an aspect of aphid physiology that needs further study. Interesting from an ecological point of view is that most of the phloem sap ingested by aphids is excreted in liquid form as honeydew. When aphids are abundant, this is produced in such large quantities that infested plants and vehicles parked beneath trees rapidly become coated with a sticky film of honeydew. Much of the sugar in this honeydew is utilized by the microorganisms that live on plant surfaces, and that which is washed by the rain into the soil is used by soil microbes.

Symbionts

Like most other insects that live on a nutritionally unbalanced diet, aphids host symbiotic microorganisms. These are housed in the cells of a structure called a bacteriosome, which is positioned on either side of the gut and immediately below the reproductive organs. The main or primary symbiont in most aphids is a bacterium, *Buchneria aphidicola*. This is closely related to *Escherichia coli* and other intestinal bacteria, including those of aphids, so it is likely that the ancestor of *B. aphidicola* was a gut microbe. Molecular and fossil evidence indicates that this symbiont was acquired some 160–280 million years ago, early in the evolution of aphids. In addition, there are bacteria and fungi which are facultative endosymbionts of aphids. That is, unlike *B. aphidicola*, endosymbiosis is not their usual way of life.

During the development of aphid embryos within their mothers, primary symbionts pass from the mother's bacteriosome and facultative symbionts from the mother's blood directly into the embryos. Evidently primary symbionts are essential for aphid growth; they are most abundant when aphids are growing and reproducing, and when aphids cease reproduction they decrease in number markedly until virtually none remain. Physiological studies indicate that they upgrade the non-essential amino acids present in phloem sap to those essential for aphid growth, and recycle nitrogenous waste. So although their food, the sole source of nitrogen for aphids, consists of a weak solution of mainly non-essential amino acids, aphids can nevertheless maintain a very high rate of increase by processing large quantities and efficiently using the amino nitrogen they extract. The symbionts undoubtedly play an important role in this efficient use of nitrogen.

In addition to their nutritional role, some scientists think symbionts are important in the ecology of aphids, in particular determining key life history traits such as defence, host specificity and reproductive activity. There is increasing evidence that the facultative symbionts affect an aphid's ability to defend itself against natural enemies such as fungal pathogens and parasitoids. For example, transferring the bacterial facultative endosymbiont *Regiella insecticola* from a clone of pea aphid that has this endosymbiont to one that has not results in the recipient clone becoming more resistant to the fungal pathogen *Pandora (Erynia) neoaphidis* (Scarborough and others, 2005)*. There is also some evidence that these facultative symbionts can improve the performance of aphids on unfavourable host plants. The symbionts, like mitochondria, are only inherited from the mother. Therefore, it is possible to make reciprocal crosses between distinct host races of aphids and obtain individuals that are similar genetically but have the symbionts of the host race associated with a particular host plant. However, the performance of these aphids on the host plants of the parents is not associated with whether they have the symbionts of the host race associated with the particular plant they are being reared on (Knäbe & Dixon, 2004). In addition, a recent molecular study indicates that, unlike in their closest free living relations, no chromosome rearrangements or gene acquisitions have occurred in the genomes of the *B. aphidicola* symbiont of aphids over the last 50 to 70 million years. Thus there is good experimental and molecular evidence to indicate that the main aphid symbiont is no longer a source of ecological innovation for its hosts (Tamas and others, 2002). That is, it is most likely that plant specialization is largely determined by nuclear genes.

The symbiosis is likely to have costs as well as benefits and thus it is advantageous for aphids to control both whether they will host symbionts and how many. Evidence from embryological studies and experimental manipulation indicates that aphids do control whether or not they host symbionts. This is another aspect of aphid biology that needs further rigorous study.

Host specificity

Most aphids are highly host specific, living on only one or two species of a genus of plants. Although there are species that live on mosses, ferns and conifers, most (90%) live on flowering plants, which make up 95% of the flora of the world. Interestingly, of the non-flowering plants the 400 species of conifers are quite exceptional in hosting 363 species of aphid. No other major group of plants hosts so many species of aphid per species of plant. Overall there is no correlation between the species diversity of the various plant groups and the number of species of aphids they host. This is surprising as several groups of aphids appear to have speciated in parallel with particular groups of plants (Börner, 1939). However, within aphid genera there are often species

* References cited under authors' names in the text appear in full in References and Further Reading on p. 128

that live on plants that belong to a group of plants other than the one that hosts the majority of the species. The implication is that aphids have not only evolved in parallel with their host plants but have also frequently acquired novel host plants (Eastop, 1972). This common and apparently casual acquisition of novel hosts is an important feature of aphids.

What is the basis of their host-plant specificity? One might expect aphids to choose to feed on those plant species on which they do best, and in a few species of aphid adaptation to living on one host is associated with poor performance on another; there is some evidence for trade-offs between performance on different hosts. But this is not always so; there are many examples of aphids doing worse on their preferred host than on other hosts. Forty years ago, Ehrlich & Raven (1964) postulated that the evolution in plants of poisonous substances (secondary plant substances) and the stepwise evolutionary responses to these by insects that feed on plants (phytophagous insects) have been the dominant factors in the evolution of butterflies and other phytophagous insects. Although this notion was not new it gained immediate and broad acceptance. In butterflies and other plant-feeding insects, the evolution of host specificity involves stepwise evolutionary responses to secondary plant chemicals; the plant acquires a chemical deterrent or toxin, and the insect evolves an ability to tolerate it, or even responds positively to it. But such stepwise evolutionary responses may not be the all-important determinant of host specificity in aphids, because most host-alternating aphids (p. 10) seasonally move between woody and herbaceous host plants which are taxonomically unrelated, for example between Salicaceae and Apiaceae (previously known as Umbelliferae), or between Rosaceae and grasses (Poaceae).

It has been argued that the great risk associated with dispersal is an important factor in the host specificity of aphids. In this case selection would favour host specificity because host plants serve not only as habitats and a source of food but also as a rendezvous for the sexes. If mating occurs on the host plant and the sexes find each other by first finding a host the individuals that settle on uninfested hosts have little chance of mating. This means that selection favours genes for extreme specificity on whichever host is commonly colonized, even if other hosts are more suitable in other respects. A similar explanation has been proposed for host site specificity in monogenean flukes and host specificity in flower mites, which use bees and hummingbirds to transport them from flower to flower. However, many physiological and behavioural experiments indicate that male aphids locate females by responding to the sex pheromone they produce from glands on their hind legs. When sexually mature the females call for males by raising their hind legs in the air and releasing the fragrant sex pheromone. Often two closely related species (p. 104) utilize the same host plant. In this case hybridization is avoided as each species produces a slightly different odour and is sexually active at different times of the day. It is often implied that males can orientate to females over long distances. If so, then the rendezvous host hypothesis cannot account for host specificity in aphids. As the mechanism by which aphids locate their mates in the field is an ecological problem it can only be resolved by field experiments. One such experiment would be to use a host-alternating aphid in which the sexes return separately to the primary host. The sexual females on cuttings of their primary host inserted into glass vials containing water can be placed into the canopies of their host plant and of a non-host plant with a similar structure and the probability of their being found by their winged males recorded. This will indicate whether the males are orientating directly to the odours emitted by the sexual females, or first to the smell of the plant and then to that of the sexual females.

Further developments in our understanding of aphid host plant associations will depend on a better understanding of the phylogeny of aphids and plants, and in particular how males locate females in the field.

Reproduction

The prodigious rate of increase of aphids has fascinated entomologists for centuries. Réaumur (1737), like Leeuwenhoek before him, thought aphids possess both male and female sex organs and are hermaphrodite. Bonnet (1745) was the first to appreciate that aphids are bisexual but can produce a succession of female offspring without males, a phenomenon that later became known as parthenogenesis, the process by which an egg develops without fertilization by sperm. Huxley (1858) was also fascinated by parthenogenesis in aphids and calculated that after ten generations, if they all survive, an aphid can give rise to a biomass equivalent in weight to 500 million stout men. Occasionally these extraordinary rates of increase are realised. Gilbert White (1887) of Selborne records that at about 3 p.m. on August 1st, 1774, showers of hop aphids fell from the sky and covered people walking in the streets and blackened vegetation where they alighted in Selborne and adjoining towns. Fortunately such plagues are rare.

Aphids give birth to live young. The advanced embryos in a mother aphid contain embryos, her granddaughters. As a parthenogenetic egg does not require fertilization, it can begin to develop as soon as it is discharged from the ovary. Under congenial conditions aphids take about a week to develop from birth to maturity, whereas other similar sized insects take approximately three weeks. However, if one takes into consideration that an aphid starts developing inside its grandmother, then the actual development time is 2.5 times longer than it takes an aphid to develop from birth to maturity, that is, approximately three weeks. Comparison of the generation times of organisms of a wide range of sizes indicates that larger organisms have much longer generation times than smaller organisms. This trend suggests that organisms the size of aphids should have generation times in the order of a month and mites in the order of a week. There appears to be a minimum 'time' required for development, which is a function of the size and complexity of organisms. Given this constraint on the rate of development, there are great advantages in the telescoping of generations seen in aphids in which parthenogenesis has enabled development to begin as soon as the egg is released from the ovary into the oviduct, and more importantly inside immature or even embryonic mothers within the grandmothers. By sliding generations inside one another aphids have reduced their generation time from three weeks to one week. In addition, because the young develop within their mothers' bodies and are born alive, aphids avoid the heavy mortality experienced in the egg stage in other insects. In this way aphids have been able to achieve the high rates of population increase normally associated with much smaller organisms like mites. This gives aphids a great reproductive advantage over other insects, and particularly over their natural enemies (p. 22).

Life cycle

Their short generation time also enables aphids to track very closely the seasonal changes in their host plants and the environment. Individuals in each generation must be able to survive the worst conditions they are likely to experience. In long-lived individuals this is likely to constrain their performance when conditions are favourable. Aphids generally are not so constrained. The seasonal sequence of short-lived generations shows generation-specific strategies, which anticipate in terms of morphology and physiology the seasonal changes in conditions (p. 14). This close tracking in time, with the matching of morphology and physiology to seasonal changes in resources, is important in determining the great abundance of many species of aphids.

In temperate regions aphids spend the winter months as eggs. In spring these hatch and give rise to nymphs that develop into adults of the first generation, known as fundatrices or stem mothers (fig. 4). They are parthenogenetic, and live bearing (viviparous); the young are born alive and develop into other parthenogenetic viviparae. Several parthenogenetic generations occur in succession until the onset of autumn when the nymphs usually develop into wingless egg-laying females and winged males. Aphids of this sexual generation mate and produce overwintering eggs. This cyclical parthenogenesis, in which periods of parthenogenetic reproduction alternate with sexual reproduction, is thought to have evolved in a seasonal climate, possibly that associated with a glacial period in the Lower Permian some 210 million years ago. However, live bearing must have evolved later as the ancestors of aphids and the closely related woolly aphids (adelgids) and phylloxerids lay eggs.

The seasonal occurrence within a species of different forms or morphs (wingless and winged parthenogenetic viviparae, males and egg-laying sexual females) is also characteristic of aphids. This is called polyphenism. Some species of aphids may have as many as eight morphs. Most deciduous tree-dwelling aphids have fewer morphs, often only winged parthenogenetic females, winged males and wingless egg-laying females (fig. 4). As the aphids that hatch from the overwintering eggs are parthenogenetic, populations are made up of groups of individuals that like twins are genetically identical (clones). In this case the evolutionary individuals on which natural selection acts are the clones. The fitness of a clone is likely to depend on the way it allocates resources to particular functions. Individuals in each clone are involved to a varying degree in defence, dispersal, reproduction and aestivation (p. 18), or hibernation. However, specialization in one or other of these functions imposes constraints on the others. At certain times of a year particular functions are more important than others for the overall fitness of a clone, and this has resulted in a division of labour within a clone, which is reflected in its polyphenism. A good example is the production by some aphids of soldier aphids. Most soldier aphids are sterile and are likely to die defending their clone mates from natural enemies (p. 26). Winged aphids similarly have a much lower fecundity than wingless aphids and a very low probability of surviving to reproduce, but in dispersing they benefit the overall fitness of their clones.

About ten percent of aphid species host alternate, spending autumn, winter and spring on a woody host and the rest of the year on a herbaceous host species belonging to a different family. Host-alternating species usually colonize herbaceous plants at a time when these are actively growing and the leaves of woody plants are mature; herbaceous and woody plants tend to show complementary growth patterns. These aphids also have more morphs than those that do not host alternate, in particular the winged spring and autumn migrants that fly from the woody to the herbaceous host(s) in spring and return to the woody host in autumn. In most host-alternating aphids the autumn migrants consist of males and the winged parthenogenetic viviparous individuals (gynoparae) that give birth to the wingless sexual females

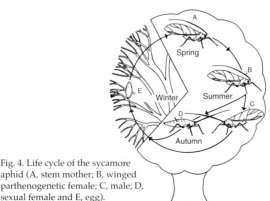

Fig. 4. Life cycle of the sycamore aphid (A, stem mother; B, winged parthenogenetic female; C, male; D, sexual female and E, egg).

(fig. 5). However, in a few species only a single winged morph is produced in autumn, which on return to the primary (woody) host quickly gives birth to wingless males and sexual females. These mature very rapidly and as they are not very mobile tend to mate with one another resulting in a high incidence of sib mating (p. 39). Many species of host-alternating aphids, like the rosy apple aphid, only colonize one species of herbaceous host during summer, whereas others, like the black bean aphid, colonize many species (fig. 6). All these species of aphid show a complete cycle, producing sexuals and laying eggs at some stage in their life cycle, and are referred to as holocyclic. There are a few anholocyclic species that reproduce continuously parthenogentically; the cycle is incomplete as they do not produce sexuals.

Host alternation was suspected over 250 years ago and first described 150 years ago, and its adaptive significance is still being debated. In spite of the high risk associated with migration from one host to another, this strategy may be adaptive mainly because clones can achieve very high rates of population increase on these two groups of plants, and so can more than compensate for the losses incurred in transferring between hosts. However, both the woody and herbaceous hosts have to be common. In the absence of plants that are common and show complementary growth patterns aphids are unlikely to host alternate.

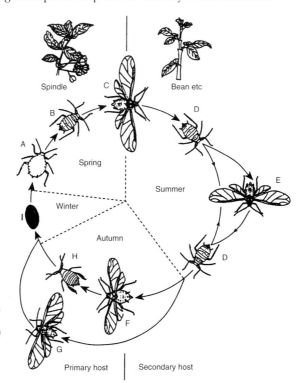

Fig. 5. Life cycle of the host-alternating black bean aphid, *Aphis fabae*. (A) Stem mother, (B & D) wingless parthenogentic females, (C) spring migrant, (E) summer migrant, (F) autumn migrant, (G) male, (H) sexual female, (I) egg.

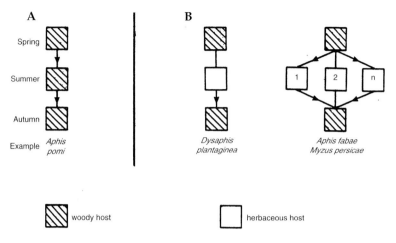

Fig. 6. Life cycles of non-host-alternating (A) and host-alternating aphids (B).

Dispersal

At all stages of development, most species of aphids move about over the surface of their host plants and even between adjacent plants. These local movements result in slow diffusive dispersal. In addition, the winged aphids also show persistent 'straightened out' movements, during which their vegetative responses are depressed, which transport them over greater distances. During these movements changes occur in the aphid's responsiveness to host plant and environmental stimuli. On taking off a winged aphid mainly responds to the wavelengths of light coming from the sky, and therefore flies upwards. After flying for a period the aphid switches from this migratory flight to alighting or targeted flight in which it responds more strongly to wavelengths of light reflected from the ground, in particular yellow wavelengths. As it stops flying before it is exhausted, it can take off again if it settles on a non-host plant. However, the number of flights is very limited as an aphid cannot survive very long without feeding. While in the migratory mode aphids are carried along by the wind as their flight speed is usually less than the windspeed. Of the aphids that migrate, fewer than one in a hundred find a host. Although it is costly for an individual, dispersal enables clones to spread the chance of survival in space and to locate plants that are of above average quality. Although the speed and direction of flight are usually determined by the wind, aphids nevertheless terminate a flight by actively flying downwards and thus have some control over the distance they travel, which in most cases is relatively short. The aphids found by Elton in 1925 on a glacier on Spitzbergen are thought to have flown there from the Kola peninsula 1,300 km away. Such infrequent but impressive long-distance migrations are achieved by the accidental riding of high altitude jet streams.

Host-alternating species of aphid show a definite seasonal pattern in flight activity. The late spring and autumn peaks in flight activity represent aphids leaving and returning to woody hosts, respectively. In some species there is an additional peak of flight activity in summer when they redistribute themselves among their herbaceous hosts. The spring and autumn flights reflect the optimum times for host transfer between woody and herbaceous host plants. Late spring is the time when the foliage of woody hosts matures and that of herbaceous plants is actively growing, and in autumn the foliage of the woody hosts begins to senesce. This senescence sets a deadline, as these aphids must produce another generation and lay eggs before leaf fall.

As it is likely that all the aphids on a short-lived plant originate from a single colonist and therefore belong to the same clone, the production of migrants is cheap as it relieves the competition for resources experienced by the rest of the colony. However, the situation is more complicated on long-lived plants, like trees, which are likely to have been colonized by many clones of varying degrees of relatedness. On trees, aphids also produce either sedentary or migratory offspring, and it is interesting to speculate on the relative benefits of the two options. The incidence of these options may indicate their adaptive significance. Populations of sycamore aphid decline by up to 63% per week during late summer mainly as a result of migration, whereas populations of other species, like *Kaltenbachiella japonica*, appear to be more or less immortal with low rates of reproduction and dispersal. We still do not know why this species shows little dispersal and closely tracks the phenology of individual trees, whereas the sycamore aphid shows high levels of dispersal and appears to track the average phenology of its host.

In summary, most aphids are host-specific, especially the tree-dwelling species. Parthenogenesis has enabled aphids to telescope generations, achieve prodigious rates of increase and closely track the seasonal changes in their host plants. Their abundance, however, also depends on the way they have overcome the physiological problems posed by the poor quality of their food – phloem sap. They are able to process large quantities per unit time and extract sufficient amino nitrogen to fuel their very high rates of growth. In this their symbionts play an important role in recycling and upgrading the quality of the amino nitrogen present in their food. The relatively few aphids that host alternate track the changes in host plant quality during a season by spending autumn, winter and spring on a woody plant and summer on an herbaceous plant. In spite of the high risk associated with transferring between hosts this strategy is adaptive mainly because clones, in achieving very high rates of population increase on these two groups of plants, can more than compensate for the losses incurred in transferring between hosts. However, the success of host alternation depends on both the woody and herbaceous hosts being common, and also showing complementary growth patterns. Although aphids are poor fliers they are nevertheless very good at riding the winds, and are occasionally transported over considerable distances; and they are very effective at locating plants even though they tend to settle on plants at random.

3 Trees as a habitat

This chapter deals with the relationship of aphids to trees, which serve as a home and source of food. In deciduous forests there can be one or several dominant species of tree, each of which hosts one or more species of aphid, most of which are highly host specific. The niche of these aphids is that of a relatively small herbivore feeding mainly on the leaves, which make up the forest canopy, one of the least explored zones on land. Because of their height trees are difficult to sample. However, their geometry can be defined and used to estimate aphid abundance and the contribution of the aphids to energy and nutrient fluxes in forests. For example, it is possible to determine the relationships between circumference of the trunk at ground level and height to the canopy area and number of leaves for a sample of trees of different sizes, and then use these relationships to estimate canopy area and numbers of leaves on trees growing in the field.

Seasonality

From bud burst to leaf fall each year tree canopies appear to change very little. However, there are changes that have shaped the life history strategies of the aphids that live there. At bud burst in spring, leaf growth is sustained by amino nitrogen imported into the leaves from the trunk. In autumn the reverse process occurs when some of the amino nitrogen is salvaged from the leaves prior to leaf fall. This flow of amino nitrogen in and out of the leaves determines the quality of the food available to aphids at any particular time. In addition, weather tends to be warmer, and the intensity of solar radiation greater, in summer than in spring and autumn. Together these changes determine the quality of this habitat for aphids. The marked seasonality in the trend in habitat quality – high in spring and autumn and low in summer – is similar each year (fig. 7). At any one time, however, not all the leaves are equally suitable for aphids. Wind and exposure to solar radiation make the microenvironment on some leaves inhospitable for aphids. Relative to their host aphids are small and as a consequence they live in a heterogeneous environment, which varies in space and time.

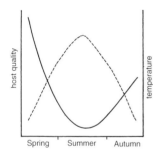

Fig. 7. Seasonal changes in host quality (solid line) and temperature (dashed line) experienced by aphids living on trees.

Surprisingly, very few tree-dwelling aphids lay their eggs close to the terminal buds. Most are laid in pores in the bark of twigs (lenticels) or crevices in the thicker older bark on the main trunk at considerable distances from the buds, especially those in the upper canopies of mature trees. A possible advantage of this is that it affords eggs protection from birds.

In many species egg hatch occurs at or shortly after bud burst. How are aphids able to synchronize egg hatch

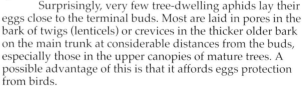

with bud burst? It appears that both aphids and trees use the accumulation of thermal units above their lower developmental threshold as a measure of time. For example, if they can only develop at temperatures above 5°C a day's exposure to 20°C is equivalent to 15 day degrees, and two days' exposure to 15°C to 20 day degrees. Although aphid and host possibly use the same mechanism for monitoring the passage of time the degree of synchronization between egg hatch and bud burst varies greatly among trees. There are very big differences in the time of bud burst between trees, and those that tend to burst their buds early in one year tend to do so every year. In this case, if it is advantageous for aphids to synchronize their development with that of their host, they should evolve a mechanism that synchronizes egg hatch with average bud burst of all the trees in an area. That some aphids have evolved such a mechanism is indicated by the close synchronization that species like the sycamore aphid have achieved between average egg hatch and average bud burst times.

What is the advantage of aphids synchronizing egg hatch with average bud burst, and what prevents them from becoming even more effective in tracking the development of their hosts? In the sycamore aphid close synchronization results in the development of large, quickly maturing and highly fecund adults; those hatching earlier or later are less fit. Few of those that hatch before bud burst survive because there is no expanding bud on which they can hide from birds or shelter from heavy rain. Those that hatch after the buds have burst take longer to develop and are smaller at maturity because they have to feed on older and less nutritious leaves. A high incidence of inter-tree movement reduces the degree to which an aphid can adapt to a particular tree, and where dispersal is frequent, as in the sycamore aphid, selection is likely to synchronize average egg hatch with average bud burst. But if dispersal is infrequent enough, natural selection will synchronize the development of aphids with that of particular trees. The scale of adaptation, fine or coarse, depends on the incidence of dispersal. Similarly, the time of leaf fall, which also varies in the same way between trees, will have determined the time of the switch from parthenogenetic to sexual reproduction and the production of overwintering eggs. Why some species show a high and others a low incidence of dispersal between trees remains to be resolved.

Interestingly, when several species of aphids live on a tree they often hatch at different times, some before and others at the time of, or after, bud burst. That is, they show marked differences in seasonal development. This is well illustrated by the aphids on birch and sycamore. The species that hatch early tend to be reproductively active early and late in a year and less active in summer. Those that hatch after bud burst tend to be active mainly in summer and to produce their sexuals in late summer or early autumn. This difference is associated with differences in the lower developmental thresholds of the species. The lower developmental threshold is the temperature below which development ceases. A low developmental threshold enables species to start developing early in a year when temperatures are still low. However, species with a low developmental threshold also have a low optimum temperature for development. At temperatures above their optimum aphids suffer heat stress. Species with low developmental thresholds and optimum temperatures often cease to grow and reproduce in summer, and enter a resting stage, either as an adult as in the birch (*Euceraphis punctipennis*) and sycamore (*Drepanosiphum platanoidis*) aphids, or as a first instar nymph as in *Periphyllus*. In addition, the sycamore aphid vacates the upper canopy and aggregates in the lower canopy of trees in summer, where temperatures can be as much as 10°C lower in the middle of the day. A few species, like *Dysaphis devecta* on apple (p. 58, 94), even have very truncated life cycles and are only active for a short period in spring, completing 2–4 generations before producing sexuals and laying eggs. So there are several ways in which aphids can exploit a particular host plant.

The role of interspecific competition in structuring community organization is an important issue in ecology. However, studies on phytophagous insects have generally cast doubt on its importance (Strong and others, 1984). In the absence of evidence for interspecific competition in natural conditions this doubt is justified. However, for each species other species are an important component of the environment. If an aphid is adapted to a particular niche, then the effect of interspecific competition on its fitness could prevent a niche shift (Akimoto, 1988). This is particularly so when the resource is a tree, which because of its long life and abundance is likely to be consistently infested with several species of aphid.

The few studies that have been done on interspecific competition in natural habitats are consistent in reporting the coexistence of aphids. The aphids living on a particular species of plant appear to partition the resources by plant part and this is associated with differences either in the morphology of the different species of aphid (stylet length and body size, pp. 3–4), their physiology (tolerance of secondary plant chemicals, p. 28) and/or their behaviour (hatching times and seasonal development, p. 14) (Hajek & Dahlsten, 1986; Akimoto, 1988; Völkl, 1989; Inbar & Wool, 1995; Jackson & Dixon, 1996). Predation and interplant movement by aphids are also thought to influence community structure (Edson, 1985). The short survival times of aphid colonies are also thought to prevent aphid numbers increasing to levels at which interspecific competition is likely to become important (Antolin & Addicott, 1988).

An interesting example of interspecific competition involves the yellow pecan aphid, *Monelliopsis pecanis*, and the black-margined aphid, *Monellia caryella*, which feed on different parts of the leaves of pecan. Leaves previously fed on by large numbers of either species of aphid have an adverse effect on the population growth rate of the other

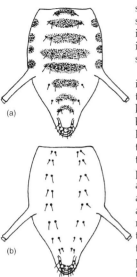

Fig. 8. Pattern of dark (melanic) pigmentation on the upper surface of the abdomen of sycamore aphids maturing in spring (a) and summer (b).

Fig. 9. Melanic patterning on the body and wings of *Lachnochaitophorus obscurus*.

species, but this effect is less than the effect on the same species (Bumroongsook & Harris, 1992). That is, the intraspecific effect on population increase is greater than the interspecific effect, and this results in coexistence. More studies of this kind are needed.

As in other insects the rate of reproduction generally increases with temperature. This means that when temperatures are low in spring and autumn, even though host quality is high, the rate of reproduction can be limited by the low temperature. At these times of the year aphids, like the sycamore aphid, tend to be much darker in colour than they are in summer (fig. 8). This is because low temperatures induce the production of dark (melanic) pigment, which enables the aphid to absorb solar radiation and maintain body temperatures as much as 2°C higher than ambient. Many of the winged morphs of tree-dwelling aphids also have dark patterning on their wings (fig. 9). In most insects such patterns are associated with mating but these particular morphs in aphids do not mate. It is likely that the melanic patterning on the wings acts like a solar panel and similarly enables them to maintain a higher body temperature and rate of increase than one would expect at the ambient temperatures prevailing in spring and autumn. When conditions become too hot aphids initially attempt to cool down by a process similar to panting. They repeatedly excrete a droplet of honeydew but instead of ejecting it hold it on the end of the abdomen for a short period before taking it back into the rectum. The assumption is that evaporation of water from the large surface area of the droplet reduces its temperature, which in turn, when the droplet is withdrawn into the body, cools the aphid. When hot they will also abandon parts of plants fully exposed to the sun and seek out more shaded areas. In this way aphids are able to a certain extent to escape the limitations imposed by temperature.

The rate of reproduction of aphids on trees also changes markedly during the season, in association with the level of amino nitrogen in the leaves of their host trees. Reproduction is rapid when the leaves are actively growing or senescing, and is slow, or ceases altogether, when the leaves are mature. Associated with these changes in reproductive rate there are changes in the size of the aphids. The adults are two to four times heavier in spring and autumn than they are in summer. Under field conditions it is difficult to separate the roles of nutrition, temperature and competition between aphids for food in determining body size. Rearing aphids in isolation or in crowds, in the laboratory, at constant temperatures, on leaves of trees that are growing, mature or senescing, results in adults of different sizes comparable to those observed in the field. Aphids reared in crowds are always smaller than those reared in isolation on leaves of the same plants. Thus both the degree of crowding experienced during development and the nutritive quality of the leaves affect adult size. Large aphids give birth to more and larger offspring and are less likely to die before reproducing than small aphids. So the seasonal changes in the nutritional quality of the leaves of trees have a marked effect on the reproductive rate of the aphids. Good nutrition results in large high quality aphids with a high reproductive potential.

The second-generation sycamore aphids, which develop and are present in summer, delay reproducing until late summer. This delay is referred to as reproductive diapause or aestivation and was thought to be a consequence of developing on mature leaves. However, when second-generation aphids are reared on young growing leaves they are large but nevertheless have well-developed fat bodies and poorly-developed reproductive organs as if about to enter aestivation. Evidently this species has a generation-specific strategy by which it is able to anticipate the conditions associated with the cessation of growth of the leaves and the high temperatures that occur in summer. The duration of aestivation depends on how abundant the aphid is, and on temperature. These aphids also differ from those in other generations in the size of their reproductive organs, length of their legs and gut, and flight behaviour. All these features are likely to be important in determining the aphids' chances of surviving the adverse conditions prevailing in summer. The other common aphid on sycamore, *Periphyllus acericola*, which hatches and completes one generation before bud burst, has a similar life history strategy but aestivates as a first instar nymph rather than an adult.

In autumn, mainly in response to short days and lower temperatures, aphids switch to producing sexual forms, which produce overwintering eggs. In this way they anticipate leaf fall by switching to a resting form prior to the onset of winter. Complementing the direct response to environmental cues, aphids have an internal clock, which governs the intensity of the response to environmental cues. Later generations give birth to proportionally more sexual offspring than earlier generations when reared at a particular day length and temperature. This clock prevents the production of sexuals in spring when days are as short as they are in autumn. The incomplete inhibition of the internal clock in early autumn enables clones to continue producing some parthenogenetic individuals at a time when the senescent foliage of trees provides a rich source of food. This is advantageous when aphids are on trees that shed their leaves late, or in autumns when leaf fall generally is late. So aphids have solved a complex optimization problem: when and how to switch to sexual reproduction in order to maximize the production of both viviparous parthenogenetic and sexual individuals.

Trees' commitment to defence

Because they are small, and most do not damage the foliage of trees, aphids are not usually thought of as adversely affecting tree growth and reproduction. However, what they lack in size they make up for in abundance. The 116,000 leaves of a 20-m sycamore tree can be infested with as many as 2.25 million aphids, equivalent in mass to a large rabbit, and the 58,000 leaves of a 12-m lime tree with 1.1 million aphids, equivalent to six great tits or four sparrows. In addition, these biomasses are turned over several times during the course of a season.

Because of the low quality of phloem sap aphids have to process very large quantities of sap in order to obtain sufficient amino nitrogen to sustain their very high rates of growth (p. 5). In the case of the giant willow aphid, *Tuberolachnus salignus*, a single aphid consumes the photosynthetic product of 5–20 cm^2 of leaf per day. The annual drain imposed on a lime tree by the lime aphid, *Eucallipterus tiliae*, is considerable. Most of this falls to the ground as honeydew, which is equivalent in energy terms to about 80% of that locked up in the leaves at leaf fall. In the case of sycamore this drain is even greater, on average equalling the energy in the leaves at leaf fall.

Sycamore saplings that are aphid infested clearly grow markedly less than uninfested saplings. Although aphids do not affect the number of leaves, when heavily infested in spring sycamore produces smaller leaves. However, the leaf area equivalent to the energy removed by sycamore aphids only accounts for a small proportion of the observed diminution in leaf area. If the drain imposed is expressed in terms of nitrogen rather than energy then the aphids again remove far less nitrogen than expected from the reduced size of the leaves. This implies that the effect aphids have on tree growth is not only a direct consequence of the energy and nutrient drain. There may be other effects. For example, the saliva aphids inject into plants may contain physiologically active components (p. 4) that adversely affect tree growth.

The width of the annual growth rings laid down in the trunks of sycamore in a year is positively correlated with the average size of the leaves, and negatively with the number of aphids on a tree throughout the year (fig. 10). This is possibly associated with the fact that each annual ring is composed of two types of vessel, which make up the spring and summer wood. The springwood is mainly laid down when the leaves are developing in spring, whereas the summerwood, which makes up most of each annual ring, is mainly laid down after the leaves stop growing. In the absence of aphids, some sycamore trees could produce as much as 280% more wood (Dixon, 2005).

The number and quality of seeds produced by plants is used as a measure of genetic fitness. The fact that many trees fruit or seed, not equally every year, but at irregular intervals, makes any study of the effect of aphids on seed production difficult. However, there are reports in the literature of aphids negatively affecting both the yield and size of seed produced by trees. In the case of sycamore, seed production over a period of eight years is not correlated with aphid numbers in either current or previous years, but is

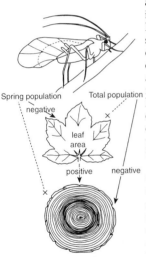

Fig. 10. Diagrammatic representation of the effect of sycamore aphid numbers on leaf area and width of the annual rings of sycamore.

strikingly dominated by the infrequent good years when seed production can be four times the long term average. Heavy aphid infestations at the time of flowering, however, can cause flowers to abort, resulting in the production of relatively few but large seeds. So by reducing the seed crop in such years aphids could have a positive effect because seed quality is greatly increased.

In the sugar maple, *Acer saccharum*, which also produces a variable seed crop, there is a strong density-dependent relationship between the proportion of seeds surviving and the number shed; when larger numbers are shed, a smaller proportion survive. Sycamore seeds germinate early and the seedlings are common beneath and around sycamore trees, occasionally forming dense carpets in woodlands. However, as soon as the buds of the trees burst and the canopy closes most of these seedlings die. This appears to be mainly a consequence of their being in the shade. It is likely that survival is dependent more on where seedlings happen by chance to be located on the woodland floor than on the number of seedlings. Therefore, it is unlikely that aphids affect the number of trees recruited. It is more likely, however, that aphids through their effect on the growth of saplings determine which will survive intra- and interspecific competition and become mature trees.

In view of the effect aphids have on tree growth one would expect trees to be selected for resistance to aphids. A detailed long term study of eight sycamore trees revealed marked differences between the trees in the average level of infestation by aphids. However, in all cases these differences could be attributed mainly to differences in the seasonal development of the trees; none of the trees appeared to show any chemically-based resistance to aphids. However, this needs to be checked by studying a much larger number of trees with the specific objective of identifying resistance to aphids.

Sycamore is a maple and in common with other species of *Acer* can have brightly coloured leaves in some autumns. This led to the suggestion by Hamilton that the red and yellow leaf colours of trees dramatically displayed in certain parts of the world signal a tree's ability to defend itself against insects. That is, the bright autumn coloration serves to signal defensive commitment against autumn-colonizing insects. Better defended trees would have brighter autumn leaf coloration.

This appears to be supported by the observation that autumnal migrants of *Periphyllus californiensis* prefer to colonize *Acer palmatum* with yellow-orange leaves over those with red leaves (Furuta, 1986, 1990). However, these trees were not all equally exposed to the sun. The autumnal colour of those in the shade is yellow-orange and those in the sun are red. More importantly, the trees in the shade shed their leaves later and burst their buds earlier than those in the sun. By colonizing trees with yellow-orange foliage in autumn the aphid is more likely to produce overwintering eggs and complete an extra generation the following spring, compared with aphids that colonize trees with red foliage. So by avoiding trees with red foliage the aphid is indeed likely to be fitter. However, what happens in late spring and early summer is equally important. The late-breaking trees, those with red foliage the previous autumn, have growing leaves, which are very attractive to aphids at the time when the leaves of the early-breaking trees are maturing and the emigrants produced on these trees then colonize the leaves of the late-breaking trees. So the trees with red foliage in autumn are likely to be more heavily infested with aphids in late spring and early summer than the trees that had yellow-orange foliage the previous autumn. By dispersing between trees the aphid exploits the spatial heterogeneity in plant quality and colonizes those trees that are the most suitable for reproduction at a particular time.

Viewed in terms of the whole season, rather than just autumn, this aphid appears to respond to cues which enable it to occupy at any particular time the most favourable part of what is a very heterogeneous environment. Being yellow-orange rather than red in autumn signals to the aphid, not a lack of defensive commitment, but that the tree's phenology is more closely synchronized with the aphid's life-cycle requirements. At other times, another part of the environment is more favourable, and in late spring and early summer it is trees that have red foliage in autumn that are preferred. Accepting this explanation leaves the question of the function of autumn coloration in trees unanswered. Botanists think the red pigment acts as a sun screen and protects the trees' diminished photosynthetic capacity from damaging radiation at a time when trees are under stress due to the low autumn temperatures. So although the idea that deciduous trees signal level of commitment to defence against aphids by the colour of their leaves in autumn is attractive and continues to be popular, there are facts relating to aphid ecology that are at odds with this theory. This idea needs to be rigorously tested experimentally.

In summary, tree-dwelling aphids live in a very seasonal and coarse-grained environment, which is nutritionally very favourable in spring and autumn but less so in summer. Each generation of aphids lives for only a short period and shows generation-specific strategies, enabling the aphids to track closely the seasonal development of their host. However, adapting to cool conditions in spring and autumn constrains their ability to develop and reproduce in summer. Synchronization of egg hatch with bud burst and sexual reproduction with leaf fall is clearly advantageous for some aphids, but if there is much movement between trees their response can only be to average time of bud burst and leaf fall rather than that for a particular tree. Although aphids have a marked negative effect on the growth and seeding of trees there is no experimental evidence to support the popular idea that trees signal their level of commitment to defence against aphids by the colour of their leaves.

4 Natural enemies

An abundance of aphids generally attracts large numbers of a diverse array of natural enemies, mainly other insects, which are dependent on aphids for their survival and therefore belong to the guild of the aphid eaters or aphidophaga. Other books in this series give accounts of particular aphidophaga such as hoverflies (Gilbert, 1993), or ladybirds (Majerus & Kearns, 1989), or aphid predators in general (Rotheray, 1989). Aphids are soft-bodied insects and appear to be very vulnerable to attack. However, they all possess a variety of behavioural, chemical and morphological defences, and in some cases can even kill their natural enemies.

Natural enemies of aphids

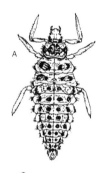

The insect natural enemies of aphids belong to one of two groups: parasitoids and predators. The parasitoids are small wasp-like insects that use a stinging action to insert an egg into an aphid, but unlike most parasites eventually kill the host (fig. 11). The larva that hatches from the egg consumes the living aphid from the inside. The adult parasitoids are free living and their food source is usually nectar or honeydew, which they use to fuel their search for aphids. Some species feed on the body fluids of aphids. In this case the aphids are first stung but no egg is inserted. The parasitoid then feeds on the body fluids that exude from the wound and this food is used both for maturing eggs and as fuel for searching. Parasitoids mainly insert their eggs into very young aphids. By the time the parasitoid larva completes its development the aphid is usually mature. Initially the parasitoid larva feeds on the less vital tissues, only feeding on vital tissues and killing the aphid right at the end of its development, which it completes within one aphid. At this stage all that remains of the aphid is its dried exoskeleton. The parasitoid larva attaches this to the substrate by means of silk and in most cases then lines the inside with silk to form a cocoon within which it pupates. This structure is called a mummy. The adult emerges from the mummy by biting a circular hole in the cocoon. These natural enemies are themselves attacked by other wasps, called hyperparasitoids, some of which are highly host specific. They are often responsible for killing a very high percentage of the primary parasitoids.

Fig. 11. A wasp (*Trioxys* species) parasitizing an aphid.

The characteristic predators of aphids are ladybirds and the larvae of hoverflies (fig. 12). These insects are free living in both the adult and larval stages. They pursue their prey, subdue it and kill it immediately on capture, and each larva needs to consume a number of prey individuals to gain sufficient food to complete its development. The survival of the very young predators is dependent on the abundance of young aphids; as with the parasitoids, the availability of

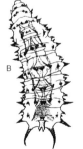

Fig.12. Larvae of a ladybird (A) and a hoverfly (B)

young aphids determines their foraging success. Like the parasitoids the predators are also attacked by specific parasitoids. In the UK the most frequently recorded are the wasp *Perilitus coccinellae*, which is a parasitoid of adult ladybirds and emerges and spins a cocoon beneath the moribund body of its host, and the phorid flies *Phalacrotophora fasciata* and *P. berolinensis*, which parasitize the pupae of ladybirds, with several individuals developing in each pupa (fig. 13).

Structure of aphidophagous guilds

Aphidophagous guilds are structured spatially and temporally, in a way that is mainly determined by habitat and the life history strategies of the natural enemies. For example, in woodland the main predator of the spindle aphid, *Aphis fabae*, is the seven spot and in parkland the two spot ladybird. Both hoverfly and ladybird larvae can regularly be found feeding on the sycamore aphid, *Drepanosiphum platanoidis*. The spring peak in abundance of young aphids on sycamore is first attacked by hoverfly larvae and then by ladybird larvae, whereas the autumn peak, which is often considerably bigger than that recorded in spring, is only attacked by hoverfly larvae. This is because hoverflies are better adapted to the cool conditions prevailing in early spring and autumn than are ladybirds. Often more than one species of ladybird will attack a colony of aphids, with the smaller of the species attacking first. Theory based on the geometry and physiology of ladybirds predicts that a large species, like the seven spot ladybird (weighing 35 mg), requires 1.5 times more aphids per unit area before it will lay eggs than a small species, like the two spot ladybird (10 mg). So the temporal pattern in the ladybird attack sequence is possibly mainly determined by geometrical and physiological constraints associated with body size, with small species of ladybird able to lay eggs at lower aphid population densities than large species.

The numbers of aphids on a plant tend to increase and then decrease, whether or not they are attacked by natural enemies. The time for which aphids are abundant on a plant is approximately equal to the time it takes the insect natural enemies to complete their development from egg to adult. Therefore, it is advantageous for the predators to lay their eggs early in the development of a population of aphids on a plant and so ensure that there are sufficient aphids to sustain the development of their larvae. This has resulted in what is referred to as the 'egg window', a period early in the increase in aphid abundance when the various natural enemies lay their eggs. Initially the food resource consists mainly of aphids but with the decline in abundance of the aphids and increase in the size of the predator larvae feeding on the aphids it changes, in terms of biomass, from predominantly aphid to predominantly natural enemy. That is, late on in the development of an aphid colony there is still sufficient food to

Fig.13. Phorid parasitoid of the pupae of ladybirds (A) and wasp parasitoid of adult ladybirds above an adult ladybird, below which is the cocoon from which it emerged (B).

sustain a predator provided it can exploit other natural enemies as well as aphids. The Asiatic harlequin ladybird, *Harmonia axyridis*, appears to have adapted to this resource, attacking aphid colonies late in their development when there are often large numbers of natural enemies present; it tends to lay its eggs later than the other ladybirds and is more efficient than they are at capturing and thriving on a diet of larvae of other species of ladybird. Interestingly this ladybird thrives less well on a diet of aphids than the other ladybirds (Dixon, 2000). Evidently specialization on a particular diet has costs in terms of performance on other diets.

This species has recently arrived in the UK and has been portrayed by the media as a great threat to our native ladybirds. The fact that one study has shown that *H. axyridis* has a very strong negative density dependent effect (p. 30) on its own abundance suggests that it is unlikely to have a significant impact on the abundance of other species. The introduction of the ladybird *Coccinella septempunctata* into North America was similarly regarded as a threat to native species. However, a recent study indicates that, as in other parts of the world, this species tends only to displace native ladybirds from agricultural crops, one of the seven spot's preferred habitats, especially when pest aphids are scarce. There is no good evidence that either of these species can displace native ladybirds from their preferred habitats (Calunga-Garcia & Gage, 1999; Dixon, 2000; Evans, 2004).

In jointly exploiting a patch of prey the members of aphidophagous guilds may affect each other's foraging success. However, the interactions between harlequin ladybirds and other aphid predators differ from competition in that one predatory species obtains food by eating another predator, a member of the same guild, and from classical predation in that it reduces potential competition for aphid prey. When one species in a guild feeds on another, this is referred to as intraguild predation. The aggressor is the intraguild predator, the victim is the intraguild prey and the common resource, the aphids, are the extraguild prey. Like the aphids in a colony, the immature stages of the various natural enemies are also at risk of being eaten. Therefore, one would expect them to be able to protect themselves in various ways. The larvae of parasitoids living within the bodies of aphids are at great risk and many are consumed along with aphids by the larvae of predators. When the parasitoid larva pupates and anchors the carcass of the aphid to the substrate then the risk increases and there are some predators, like anthocorid bugs, that depend on 'mummified' aphids to complete their development. The documentation of the interactions between the larvae of the various species of ladybird indicates that there is great variation in the nature of their chemical, morphological and behavioural defences. The larvae of some species, like the seven spot, tend to drop off a plant when they encounter the larva of another species, especially if it is bigger. Others, like the two spot, are particularly well defended chemically and if eaten can adversely affect the development and survival of the predator.

Aphid escape and defence mechanisms

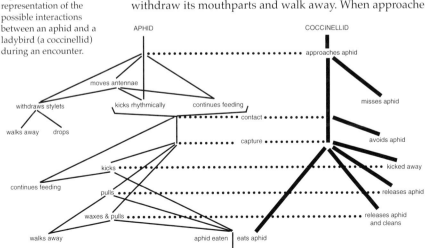

That aphids can defend themselves has been known since 1891, when Büsgen described 'waxing'; several species of aphid exude an oily liquid from their siphunculi (p. 1; fig. 14), smear this on to the head of an attacking predator and then escape. Other researchers have observed that some aphids, when disturbed by parasitoids and predators, escape by leaping with alacrity from their host plant.

Aphids feeding on the stems of plants generally face downwards and those on leaves face towards petioles. As the larvae of insect predators tend to move up a stem and on to leaves via the petiole, this behaviour is advantageous as aphids can see predators approaching. This is supported by the observation that aphids have a greater chance of avoiding capture by a predator when it approaches from the front than from the rear. Thus, the position of the siphunculi on the rear part of an aphid's abdomen is possibly advantageous in that they are in the most appropriate position to facilitate escape once the aphid has been caught.

Pursuing, subduing and eating prey can be risky for ladybirds. The responses shown by aphids and insect predators to each other are shown in fig. 15. The relative sizes of the aphid and predator determine to a great extent the outcome of an encounter. Aphids can avoid capture by kicking, walking away or dropping off a plant when encountered by a predator. While it is advantageous for aphids to repel predators and continue feeding, they can only do this if the predator is much smaller than they are. If the predator is not deterred by kicking and continues to approach, an aphid's most effective means of escape is to withdraw its mouthparts and walk away. When approached

Fig. 14. Aphid 'waxing', daubing siphuncular exudate onto the head of a ladybird larva that has captured it by one of its hind legs.

Fig. 15. Diagrammatic representation of the possible interactions between an aphid and a ladybird (a coccinellid) during an encounter.

by a large predator, which can walk faster than even an adult aphid, then it is expedient for the aphid to drop or jump off a plant, although it might die of starvation before it is able to find another host plant. So if it does not avoid the predator it will be eaten and if it does it may starve to death. Similarly, in the case of the predator, if it is small relative to an aphid it is approaching it risks being kicked off the plant, which could result in death, or aphids may daub so much siphuncular exudate onto them that they cannot move and starve to death. The balance of probabilities has shaped the responses of both aphids and predators.

If a predator seizes the leg or antenna of an aphid the victim may attempt to kick the predator away, or pull the appendage free. If the predator is small relative to the aphid then the predator can often be kicked off a plant. In some aphids, if pulling fails, the aphid sheds the leg held by the predator or daubs secretion from its flexible siphunculi on the predator's head (fig. 14). The predator then appears to become preoccupied with removing the siphuncular secretion and pulling is then frequently effective in enabling the aphid to escape. The siphuncular secretion is only effective when the predator and aphid are similar in size (Dixon, 1958). Hymenopterous parasitoids are also frequently entangled in the siphuncular secretion, which may be a more effective means of protection against parasitoids than against predators. When an aphid that is captured and is being eaten by a predator, or being attacked by a parasitoid, produces a droplet of secretion at the tip of each siphunculus (figs. 11 & 14), other aphids close by cease feeding and walk away in response to the smell of the secretion and the sight of the struggling aphid. In the peach–potato aphid the smell of the siphuncular secretion alone will make the other aphids walk away or drop off a plant. The substance that initiates this response is an alarm pheromone, the most common in aphids being trans-β-farnesene. By producing siphuncular secretion an aphid may secure its escape and warn others of the presence of a predator. As adjacent aphids are likely to be of the same clone, sharing its own genes, this altruistic act increases the probability of the aphid's genes surviving.

Aphids living on the exposed parts of plants generally have long flexible siphunculi whereas those protected in galls, or on roots, or which tend to be hirsute or covered with flocculent wax or attended by ants, have short or rudimentary siphunculi, which are less valuable as protective devices. Thus the degree of development of the siphunculi varies, and aphids with short siphunculi usually have other means of protection.

In some species there are individuals that are specialized for defence, the soldier aphids. They can make up 13 to 50% of a colony, are very short lived, do not feed or reproduce, and defend the colony against insect enemies (fig. 16). They are characteristically well armoured, often with robust prehensile fore limbs and frontal horns. After seizing a potential enemy with their fore limbs they impale it on

Fig. 16. Soldier of *Pseudoregma alexanderi*.

their proboscis or horns. Soldiers of some species insert their stylets into potential enemies and inject a cysteine protease, which causes paralysis and death. The venom appears to be produced in the mid-gut, which implies that the soldiers must regurgitate their gut contents. Soldiers of the sugarcane woolly aphid, *Ceratovacuna lanigera*, in Japan respond to the alarm pheromone emitted by reproductives, not by dispersing, but by aggregating and attacking any object contaminated by siphuncular exudate. As a colony is likely to be a clone, consisting mainly of the descendants of a founder aphid, soldiers reduce the probability of a clone becoming extinct. Approximately 50% of the 100,000–200,000 inhabitants of the large galls produced by the aphid *Ceratoglyphina bambusae* on snowball trees in Japan are soldiers. The soldiers live not only in the gall, but also on the gall and adjacent twigs. When a gall is handled or shaken the soldiers fall or stream off the gall and can inflict painful bites on man. These biters possibly afford a colony some protection against squirrels, which are known to eat the contents of aphid galls. The investment in soldiers fluctuates in a regular fashion within the life of a colony, seasonally, or as a density dependent response to the size of the colony or the gall. The soldiers of *Nipponaphis monzeni*, which induces galls on trees in Japan, are even capable of repairing damage caused to the wall of a young gall by caterpillars. Immediately a hole is made in the wall of a gall soldiers gather around the hole and discharge fluid from their siphunculi onto the damaged area and because many soldiers participate the volume of fluid increases rapidly. Using the legs and proboscis they plaster up the hole with the fluid, which soon becomes viscous and congeals. During this process the soldiers shrivel as they discharge the fluid and several often become buried in the plaster. So the original function of siphunculi for defence against natural enemies can be modified for plastering up holes in galls. Although the aphids that form galls and produce soldiers are mainly to be found in the Far East there are a few species of the subfamily Pemphiginae, in particular the genus *Pemphigus*, which occur in Europe and are worthy of further study (Foster & Northcott, 1994; Stern & Foster, 1996).

Aphids exhibit two main trends in the evolution of their behaviour in relation to natural enemies. Many species depend chiefly upon their activity to avoid approaching natural enemies. These aphids are characterized by cryptic coloration and often form diffuse colonies. In contrast, inactive species do not avoid approaching natural enemies and often live in association with ants (that is, they are myrmecophilous) and are conspicuous as the result of both their coloration and their tendency to form large compact colonies. Other aphids like the mealy plum aphid, *Hyalopterus pruni*, and vetch aphid, *Megoura viciae*, are distasteful and poisonous, respectively, to some species of ladybird (Dixon, 1958). In some cases the species of host plant of an aphid can determine its suitability as prey. For example, Professor Pasteels while on holiday in Greece

observed that the larvae of the Adonis' ladybird feeding on
the oleander aphid, *Aphis nerii*, living on *Cionura erecta*, gave
rise to adults that were wingless or had small non-functional
wings, whereas those that developed from larvae fed on the
same aphid living on *Cynonchum acutum* or *Nerium oleander*
had functional wings. Similarly, seven spot ladybird larvae
fed on the lupin aphid, *Macrosiphum albifrons*, from bitter
lupins, which contain high concentrations of the alkaloid
lupanine, show abnormal development or die, whereas those
fed the same aphid from sweet lupins develop normally.

Ladybird larvae can learn to avoid unpleasant tasting
aphids. Ten spot ladybird larvae which catch and feed on the
mealy plum aphid will after a short time release the aphid.
On encountering another individual of this species of aphid
they will seize it but release it before starting to feed. So they
appear to be able to learn to avoid feeding on certain prey,
reducing the risk of poisoning. It would be interesting to
know just how widespread this behaviour is. It is the adult
predators that decide where to lay their eggs, and as most
aphid colonies are made up of one species the larvae of
predators have very little choice but to eat the aphids they
encounter. Aphids that allow ants to tend them are protected
against their natural enemies, as the ants attack and remove
these from aphid colonies they are tending. In some cases the
ants will further protect aphid colonies by enclosing them in
a shelter, which they construct from soil particles. So long as
these aphids provide the ants with an abundance of
honeydew, which the ants use mainly as a source of energy
for foraging for high protein food to feed their brood, then
the ants remain with the aphids and protect them. However,
this protection is dependent on the general availability of
honeydew and high protein food for the ants. One might be
tempted to view the interaction between ants and aphids
only in terms of benefits for both parties. However, relatively
few species of aphids are ant attended. This raises the
question whether there may be costs as well as benefits for
ants and aphids in sustaining such a relationship. Very few
studies have addressed this problem. All indicate that there
are costs for the aphid in terms of its rate of growth and
reproduction. Although honeydew is a waste product,
producing it in sufficient quantity and quality to be attractive
to ants appears to be costly for aphids. In the case of
facultatively ant attended aphids it appears they have to feed
faster than is optimum for the assimilation of nutrients from
phloem sap. As a consequence they grow and reproduce
more slowly than when not ant attended. For those aphids
that are always ant attended (obligately), the association
between ant and aphid is very close and therefore very
difficult to analyse in terms of costs and benefits. Ant
attendance is not associated with specific taxonomic groups
of aphids but appears to have evolved independently in
many groups. Why some species of aphids have evolved this
mutualism with ants and others have not is still largely
unknown. Indeed the relationships between ants and aphids
need a great deal of further study (Stadler & Dixon, 2005).

In summary, when abundant, aphids attract large numbers of a diverse range of mainly insect natural enemies. Aphids have evolved an awesome array of behavioural, chemical and morphological defences against predators and parasitoids. In some species there is even a specialized non-reproductive soldier morph, which protects aphid colonies against natural enemies. Other species utilize ants to protect them, which is costly for the individuals but advantageous for the colony.

5 Abundance

As we have seen, aphids have a prodigious rate of increase (p. 8) and are potentially capable of becoming very abundant over a wide area. Fortunately such outbreaks are rare. The implication of this is that aphid abundance is normally regulated well below plague levels. Before we continue it is important to establish what processes act on a population and how they are likely to affect the level of abundance. A population can be likened to the water in a bath into which water flows from a tap and from which it exits via a pipe in the bottom of the bath (fig. 17). If the rate of water flow into the bath exceeds the rate of outflow then the water level rises. However, as the water level rises the rate of outflow increases until it equals the rate of inflow at which point the water level ceases to rise and remains at a constant level, so long as the rate of inflow does not exceed the capacity of the outflow pipe. The rate of outflow of water depends on the amount of water in the bath. Varying the diameter of the drainage hole affects the rate of outflow and the water level. However, if the rate of outflow did not vary with the depth of the water in the bath then the water level would not stop rising. This is very similar to what happens in a population, which increases in abundance (water level) until the rate of increase (water inflow) is the same as the rate of loss of individuals (water outflow). That is, regulation is dependent on the rate of loss of individuals (water outflow) increasing as the number of individuals per unit area (water depth) increases. That is, for regulation the losses from a population have to be dependent on the density with the actual level of abundance determined by the increase in the rate of loss with population density. In other words, for regulation a mortality factor has to act more severely on a large population than on a small one. This situation is referred to as density dependent regulation. Regulation can also be achieved if the density dependent effect operates not on the losses but on the rate of recruitment to the population, in which case the level at which it stabilizes depends on the rate at which recruitment is reduced with increase in population density. All other factors that affect recruitment or loss but operate independently of density may affect the abundance achieved but will not regulate the population.

We now consider the role of natural enemies, host abundance and intraspecific competition for resources as potentially important factors affecting aphid abundance.

Fig. 17. A bathtub diagrammatic representation of population dynamics. The tap controls recruitment and the plug hole controls the losses from the population, the abundance of which is represented by the level of water in the bath.

Role of natural enemies

When aphids are abundant, they are usually attended by large numbers of natural enemies of various kinds, and it is often assumed that these regulate aphid abundance. This is particularly the case for ladybirds. The prevalence of holy attributes, like our lady, in their common names in all European languages reflects the widely held belief that they are harbingers of good tidings. This notion is sustained by the current widespread use of ladybirds as an emblem implying the effectiveness of biological control and a pesticide-free product.

The assumption that natural enemies act in a density dependent way on populations and so regulate abundance is unlikely to hold for aphids because their rates of development and increase greatly exceed those of their natural enemies. In addition, the natural enemies have first to find the aphids. Once these are found, it is rash to assume that it is advantageous for natural enemies to reduce aphid numbers dramatically.

There is no doubt that natural enemies can sometimes reduce the numbers of aphids dramatically, and the use of hymenopterous parasitoids and ladybirds in biological control of aphids is claimed to have been successful. Indeed the first outstanding success in biological control was the use of a ladybird (*Rodolia cardinalis*) to control the cottony cushion scale (*Icerya purchasi*), a close relative of aphids. This success reinforced the belief that natural enemies regulate insect abundance. Indeed the array of insect natural enemies that attack the sycamore aphid is awesome (Table 1) and the four primary parasitoids only attack the sycamore aphid, that is, they are host specific.

Table 1

Parasitoids	Predators
Aphelinus thomsoni[a]	*Adalia bipunctata* (a ladybird)
Monoctonus pseudoplatani[a]	*Anthocoris confusus* (a bug)
Trioxys cirsii[a]	*Anthocoris nemorum* (a bug)
Endaphis perfidus[a]	*Chrysopa carnea* (a lacewing)
	Chrysopa ciliata (a lacewing)
	Episyrphus balteatus (a hoverfly)
	Syrphus vitripennis (a hoverfly)
	Tachydromia arrogans (an empidid fly)

[a] Host specific

In spite of the success of *Rodolia*, ladybirds have not proved effective biological control agents of aphids in the field (Dixon, 2000). The reason for this is to be found in the dynamics of the aphid–ladybird interaction and its consequences for ladybird fitness. A major factor in this is that aphids develop much faster than ladybirds (p. 9) and as a consequence aphids can disperse and become quite scarce locally before ladybird larvae can complete their development. This is important because ladybird larvae, unlike the more mobile adult ladybirds, cannot seek aphids elsewhere. Natural enemies that have a developmental time similar to that of their prey are potentially capable of regulating the abundance of their prey as is the case with the

Rodolia/Icerya interaction. *Rodolia* has a developmental time similar to that of *Icerya*. That is, the generation time ratio of ladybirds and their scale insect prey is close to 1:1.

Of the insect predators attacking the sycamore aphid, the anthocorid bugs and ladybirds, which mainly attack the spring peak in aphid abundance, have been studied in detail. They both have developmental times considerably longer than that of the aphid. The survival of the early stages of anthocorid bugs and ladybirds is very dependent on an abundance of young aphids and that of the later stages of anthocorid bugs on an abundance of parasitized aphids (mummies) and adult aphids. However, the proportion of the sycamore aphid population these predators kill decreases as the sycamore aphid population increases in abundance. Another important group of predators is the larvae of hoverflies, the syrphids. Although they have not been studied in detail, they also have relatively long developmental times, mainly attack sycamore aphid populations in early spring and autumn, and as with the previous two predators there is no indication that they act in a density dependent way. So insect predators do not appear to regulate sycamore aphid abundance.

Hymenopterous parasitoids lay eggs in aphids. These eggs give rise to larvae that feed on the internal tissues of the host (p. 22). Each parasitoid matures on one aphid and they are generally thought to take approximately the same length of time as their host to complete development. Therefore they would appear to be potentially capable of tracking and regulating aphid abundance. However, the parasitoids of sycamore aphids actually take considerably longer than the aphid to complete their development. Generally there are two peaks each year in the numbers of mummies of the two commonest parasitoids, *Aphelinus thomsoni* and *Monoctonus pseudoplatani*, one in June–July and the other in August–September. As the time it takes for these parasitoids to develop from oviposition to mummification is known it is possible to relate the number of mummies appearing to the number of young aphids present at the time of oviposition. The proportion of young aphids parasitized declines with increase in aphid abundance both early and late in a year, with the levels of parasitism late in a year much lower than those early in a year. In addition, there is often no second peak of mummies, even when the autumn peak of aphids is similar in size to those in years when there is a second peak of mummies (fig. 18). There is a very strong association between the size of the second peak of mummies of both parasitoid species and the availability of young aphids in mid-July to mid-August. If the summer reproductive diapause of the sycamore aphid is prolonged then there are few or no young aphids during this period and no second peak of mummies, but if the diapause is short there are many young aphids and a large second peak of mummies. So the size of the second peak of mummies is related, not to the autumnal abundance of the aphid, but to the presence of young aphids in late summer. This is because the parasitoids

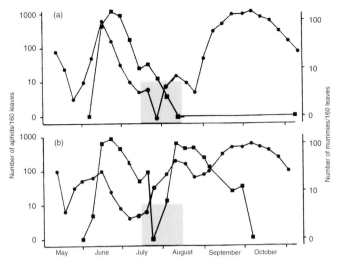

Fig. 18. The seasonal trend in the number of young aphids ●—● and number of mummies ■—■ of *Aphelinus thomsoni* in a year when there was no second peak of mummies (a) and when there was a second peak of mummies (b). The shaded block is the period during which young aphids should be abundant if there is to be a second peak.

have to complete their development before leaf fall, and ovipositing after mid-August leaves insufficient time for them to do so. In addition, the spring peak of mummies is not correlated with the peak number of mummies the previous June–July or August–September. Not only are the parasitoids not acting as density dependent regulating factors, but the dynamics of the aphid and parasitoids in this case frequently become uncoupled in autumn.

The effectiveness of the parasitoids is also reduced by the action of their own parasitoids (hyperparasitoids) and predators, which in many cases are less host specific than the primary parasitoids. In addition, because of the risk of hyperparasitism, primary parasitoids are likely to cease ovipositing when the proportion of aphids parasitized is relatively low as a high proportion of primary parasitism makes the population attractive to hyperparasitoids. By continuing to oviposit where the incidence of parasitism is high a primary parasitoid may reduce its potential fitness. As with predators, their foraging behaviour has been selected to optimize their fitness, not to maximize their effectiveness as biological control agents.

In 1970 several hymenopterous parasitoid species were introduced from Europe into America in an attempt to reduce the abundance of the lime aphid (p. 66, 98), which is regarded as a social nuisance because of the large quantities of sticky honeydew it produces. Only one of the parasitoids became established and initially was claimed to be effective, but a seven-year study could find no evidence that the parasitoid was regulating the abundance of the lime aphid. During a 20-year population census of the Turkey oak aphid (*Myzocallis boerneri*) (p. 70, 101), no parasitoid mummies were recorded. So regulation by parasitoids is certainly not a general feature of aphid population dynamics.

Summarizing, there is no field evidence that unambiguously indicates that parasitoids are capable of regulating aphid abundance, and there is also good empirical support for the theoretical prediction that in nature aphidophagous predators are ineffective regulators of aphid abundance.

Role of host abundance

As we have seen, equilibrium population density of a species is the outcome of the interaction between its rate of increase and density dependent mortality factors; and the natural enemy induced mortality of tree-dwelling aphids is not density dependent. Moreover, there is no evidence that the various natural enemies of the different species of aphid differ in efficiency. Therefore, if one assumes that the density dependent mortality factors operate similarly in all aphids, the most likely cause of the marked differences in abundance between species is differences between aphid species in their intrinsic rate of natural increase (the increase in numbers per individual per unit time), or the difference between birth rate and death rate. This effect is similar in its consequences to variation in the rate of flow of water into the bath (recruitment) (p. 30).

What factors are likely to affect the degree to which the intrinsic rate of increase is realized? One such factor is the probability of finding a host plant. In organisms like aphids, which regularly disperse between plants, have little or no control over their flight direction and tend to settle on plants at random (p. 12), the probability of finding a host plant is likely to be directly proportional to the area of ground covered by the host plant species. Thus, all other things being equal, the abundance of the host plant, through its effect on realized rate of increase, should markedly affect the abundance of an aphid. This prediction is well supported by a study of the various species of deciduous-tree-dwelling aphids. Those that live on common species of tree usually occur in abundance on those trees whereas those that live on uncommon trees are far less abundant (fig. 19). Thus, plant

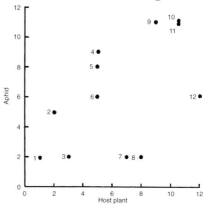

Fig. 19. The relationship between rank order of abundance of 12 species of deciduous tree-dwelling aphids and the rank order of abundance of their host trees ($rs = 0.52$, $P< 0.05$).

abundance is likely to be a major factor determining the differences in aphid abundance on different species of tree.

Interestingly there are striking exceptions to the above prediction. One such example is the aphid *Monaphis antennata* (p. 61, 100), which lives on birch. Its host is common and widely distributed, from Ireland in the west to Eastern Siberia, and from Scandinavia in the north to the Mediterranean. The aphid, although recorded throughout much of the distributional range of its host plant, is always very rare relative to the other species of aphid living on birch. There appear to be no areas where this aphid can be said to be abundant, which is surprising in view of the abundance of its host; this species of aphid is truly rare relative to its resource. Rarity intrigues biologists. A few species are very abundant but the majority are uncommon to rare. Not surprisingly *M. antennata* differs in certain features of its life history strategy from the other aphids living on birch. However, it still remains to be resolved whether the observed differences are the cause or the consequence of rarity. This must be addressed if a general theory of rarity is to be developed.

Role of intraspecific competition

If natural enemies do not regulate aphid abundance, what does? One possibility is that intraspecific competition for a limited resource could be the regulatory process. This would affect fitness in terms of the contribution of individuals to future generations. Its occurrence is usually measured in terms of survival or fecundity. The effect of intraspecific competition is density dependent in that the adverse effect of competition on the fitness of individuals is greater the greater the number of competitors.

For most species of insects there is little evidence of competition for resources. However, many field and laboratory studies on aphids have revealed intense intraspecific competition for resources. In particular a detailed study of the sycamore aphid has shown that when abundant this species competes for suitable space and resources, and the effect of this competition is strongly density dependent. It results in direct and delayed reductions in the rate of reproduction, and an increase in migration and mortality. In addition, the density-dependent processes act continuously and often in parallel with one another. This results in the sycamore aphid in optimum habitats showing a marked 'see-saw' in abundance in which high numbers in spring are followed by low numbers in autumn and *vice versa*. Evidently the abundance of this aphid is regulated by strong density-dependent intraspecific competition for resources. This auto-regulation is likely to be a general phenomenon, as similar 'see-saws' in abundance within years, associated with density-dependent recruitment and migration, are recorded for other tree-dwelling aphids.

For good pragmatic reasons, there are few long-term studies on insects. But such studies need to be encouraged as they provide the reality against which to test theoretical predictions. Population studies (p. 30) at different sites within the range of an aphid should also be encouraged, as there are indications that the within-season dynamics of the sycamore aphid differ in the north and south of the UK. The sycamore aphid thrives in the cool humid conditions prevailing in the north of the UK and does badly in the hotter and drier conditions prevailing in the south. It is possible that the populations in the south are sink populations maintained by density dependent migration of aphids from the much higher density source populations in the north.

In summary, although natural enemies are an important cause of mortality, there is no evidence that they regulate the abundance of tree-dwelling aphids. They appear to be constrained by their dependence on the availability of young aphids, their relatively slow rate of development compared to aphids, and the need to synchronize their life cycle with that of aphids. This last requirement, in particular, can result in the dynamics of the parasitoids often becoming uncoupled from those of the aphids. The abundance of aphids living on common species of trees, and the rarity of those on less common hosts, appears to be due to the losses of aphids during dispersal being dependent on host abundance. There is good evidence to indicate that aphids, when abundant, compete for suitable space and resources. This competition has strongly density dependent effects, and it is the most likely factor regulating aphid abundance.

6 Sex

As we saw in Chapter 2, aphids can switch from reproducing without males to sexual reproduction. This has enabled them to track very closely seasonal changes in their environment by evolving very complicated life cycles. Although much of the detail of their life cycles can be accounted for in terms of ecology and physiology there are still many mysteries that remain: in particular, why when reproducing sexually the ratio of females to males is always biased in favour of females.

The life cycle of most tree-dwelling aphids consists of a series of parthenogenetic generations culminating in the production of males and sexual females, which lay the eggs. In temperate regions egg-laying coincides with the onset of winter, but in other parts of the world it coincides with the onset of a hot or dry season. The switch to sex should be postponed as long as possible as it ends the phase of rapid parthenogenetic reproduction. However, mating (fig. 20) and oviposition must occur before the end of the favourable season, which, for aphids living on deciduous trees in temperate regions, is leaf fall. The switch to producing sexuals is cued mainly by short day length and low temperature as the sexual forms are environmentally induced. Environmental cues probably trigger neuroendocrine changes, which control the sex of the offspring at the time of the maturation division of the eggs. The result is that the males and females are present for a relatively short period in autumn at a time when population abundance is changing very rapidly (fig. 21). Both sexes make only a brief appearance and their relative numbers do

Fig. 20. Currant aphids (*Cryptomyzus* species) mating.

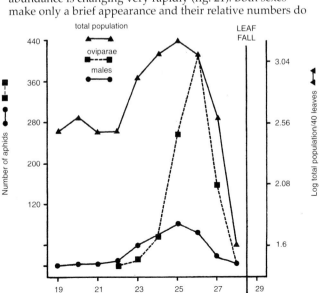

Fig. 21. Changes in the abundance of sycamore aphids that occur on the leaves of sycamore in autumn prior to leaf fall. The trends are for the logarithm of the total number of aphids, and the actual numbers of adult males and sexual females (oviparae).

not achieve equilibrium. In addition, the fitness of a male depends on the number of females it fertilizes, while that of the female depends on her probability of mating.

Deciduous tree-dwelling aphids in the temperate region live in a markedly seasonal environment, which is only suitable for reproduction up to the time of leaf fall. In those species of aphids that live on evergreen plants low temperature is likely to be the factor that brings the favourable season to an end. So on both deciduous and evergreen trees, there is some time by which aphids should complete the laying of eggs. As we saw on page 10, this presents aphid clones with an optimization problem: when and how to switch to sexual reproduction in order to maximize the production of both viviparous parthenogenetic and sexual individuals. Aphids appear to have resolved this problem, in part at least, because there is a high degree of synchronization between the appearance of sexual forms in many species and the onset of senescence of the leaves of their host tree. This poses some interesting questions. What mechanism(s) do aphids use to synchronize their phenology with that of their host? And what is the adaptive significance of the pattern in time in the allocation of resources to male and female production?

Mechanism

Aphids use day length as an indicator of the time of year; the switch to producing sexuals is triggered by short day length experienced in autumn. Although many aphids hatch from eggs early in the year and are present in spring when days are as short as they are in autumn, the sexuals rarely appear in the field before autumn. This is because of the operation of an intrinsic timing mechanism, an 'interval timer' (Lees, 1960). In the sycamore aphid it is not that the virginoparae are insensitive to the short day lengths in spring, as they can produce sexual females if experimentally exposed to very short day-lengths, shorter than those prevailing in spring. So in this species the controlling mechanism is more like a resistance that decreases with time than a clock that switches the mechanism on and off. In addition, the incomplete inhibition of the interval timer in autumn enables clones to continue to produce some parthenogenetic individuals in autumn when the senescent foliage of sycamore provides a rich source of food. This is advantageous for aphids on those trees that in any one year shed their leaves late, or in autumns when leaf fall is late. In contrast males do not appear until the fourth generation, and their production does not appear to be related to day length or temperature.

Sex ratio

The ratio of the number of sexual females to males (sex ratio) in the sycamore aphid varies between years and trees, ranging from 0.8 to 3 times as many sexual females as males. Although we do not fully understand why the sex ratio varies so much, the consistent feature is that the sex ratio is biased in favour of females. The birch aphid, *Euceraphis betulae*, has a similar female-biased sex ratio with two to three times as many sexual females as males. In other species the bias in favour of females can be even greater; for example, the sex ratio is 1:5 (males:females) in the alder aphid *Pterocallis alni*, 1:7 in the lime aphid *Eucallipterus tiliae* and 1:13 in the Turkey oak aphid *Myzocallis boerneri*. Interestingly, there is some indication that the lower the female bias in the sex ratio the more time males spend guarding their mates. When females are a relatively rare resource it is advantageous for males to guard any female they find and prevent her from mating with other males.

Fisher (1930) predicted that natural selection should drive populations to an equilibrium at which half of the parental resources are allocated to sons and half to daughters. Imagine a population with an excess of females, where a male can obtain several mates. Each male in these circumstances will then have, on average, a greater number of offspring than each female. All females produce about the same number of offspring, but a female would have more grandchildren than other individuals if she produces sons rather than daughters. If this trait is heritable then the male-producing tendency will spread in the population. If male births increase to such an extent that the sex ratio becomes male biased then it becomes more advantageous to produce daughters. Natural selection will act to favour the production of the rarer sex. The outcome is an equilibrium sex ratio of 1:1. In the case of mammals the process is currently mainly predetermined in that females have two X chromosomes and males an X and a Y chromosome. Therefore, all other things being equal, there should be on average equal numbers of males and females. The above and models for other organisms assume that the sex ratio is not only at evolutionary but also at demographic equilibrium. In organisms with brief mating seasons, like tree-dwelling aphids, however, this is not true (fig. 21) as their populations are very unstable.

Two processes have been proposed to account for biased sex allocation ratios in aphids. First, if females mate with males that are closely related to them then local mate competition results in the selection for female-biased sex allocation (Hamilton, 1967); in the extreme case – sib mating – a parent can maximize the number of grandchildren by producing many daughters and just enough sons to mate with them. This process possibly largely accounts for the markedly female-biased sex ratios observed in the Pemphiginae. In this subfamily of aphids short days in autumn induce the appearance of sexuparae, a morph that quickly gives birth to only males and sexual females. Both mature within hours of being born, and they are small and not very mobile. This means that the sons are likely to mate with the daughters, especially when the aphid is relatively uncommon. As most of the Pemphiginae host alternate, with the species returning to primary woody hosts in autumn after spending the summer on herbaceous plants, it is likely that very few sexuparae will colonize trees in certain years and particular trees in any one year. However, most tree-dwelling aphids do not host alternate, but remain on the same host throughout the year. In these species short days in autumn induce them to produce a mixture of virginoparae, males and sexual females, the proportions of which vary in time. In addition there are likely to be many clones on a tree, so that inbreeding or local mate competition is likely to be uncommon. This is particularly the case in the sycamore aphid, which compared to other tree-dwelling aphids shows very high levels of dispersal (p. 13) and consequently mixing of clones. This was confirmed by a genetic analysis of sycamore aphid population structure, which indicates that the populations on individual trees are indeed made up of many clones. In addition, the abundance of the aphid and environmental conditions change very rapidly at this time of the year. In such cases the fitness of males and females may depend in different ways on age or environmental variables. As the production of males and sexual females is environmentally induced the skew in the sex ratio probably depends on the form of the relations between fitness and age (or environment). For example, if some males produced early in autumn survive to mate with females produced later in autumn, then the evolutionarily stable strategy is a male-biased allocation early and female-biased allocation late in autumn. This is similar to the system prevailing in the sycamore aphid where reproduction is continuous and the generations overlap.

The precise predictions depend on whether females mate repeatedly and fertilize each egg with sperm from the most recent mating, or mate only once, and on the rates of search, and of male and female mortality. Field data for the sycamore aphid indicate that sexual females are just as successful at attracting mates when population density is low as when it is high, but there is some evidence that males have to compete more for mates as population density increases, as they disperse more when population density is high. However, much remains to be resolved. That the fitness of a male depends on the number of females it fertilizes, while that of a female depends on her probability of mating has two important consequences. First, if males are long lived, then there are advantages in their maturing before the females. This is known as protandry. Secondly, there should be a seasonal bias in the sex allocation, depending on how the total reproduction rate varies through the season and mainly on the longevity of males. Overall this should result in a variable but female-biased sex allocation, which is what is observed.

Both the female bias and the degree of protandry in the sycamore aphid also appear to be positively associated, significantly so, on some trees. As males are winged and more mobile than the wingless sexual females and more flight active than the winged parthenogenetic females, males may move between trees. Thus the sex ratio on any particular tree could be affected by the movement of males between trees, and their tendency to accumulate on those trees that shed their leaves last in any one year. In addition, mature sexual females spend a lot of time each day laying eggs on the trunk and main branches of their host trees and are therefore not included in the sample counts of aphids on the leaves. So sex ratios based on field counts of aphids on leaves are unlikely to be accurate. However, rearing this aphid on saplings under near natural conditions in the field in autumn and under short day conditions in the laboratory also indicates that the adult sex ratio is female biased. Therefore, although the precise operational sex ratio in the field is uncertain, the data indicate that it is variable and female biased.

In other species like the alder, lime and Turkey oak aphids the sex ratios based on leaf counts are more markedly female biased. One explanation for this is that local mate competition is more likely to occur in these species than in the more migratory sycamore aphid. It remains to be shown, however, that populations of these aphids are made up of fewer clones than is the case in the sycamore aphid.

In summary, tree-dwelling aphids, in common with other aphids, have a female-biased sex ratio and environmental sex determination. The sexes make only a brief appearance in autumn and neither achieves an equilibrium density. In addition, the fitness of a male depends on the number of females it fertilizes, while that of a female depends on the probability of mating. The fact that aphid populations are made up of clones increases the probability of females mating with their male relatives. However, the fact that populations of non-host-alternating tree-dwelling aphids are multiclonal tends to rule out local mate competition as the major factor determining the female-biased sex ratios. As reproduction is continuous and generations overlap, the evolutionarily stable strategy is for these aphids to be protandrous and have a female-biased sex ratio. This is probably mainly because the sexes only make a brief appearance, they do not achieve equilibrium densities, and each has a different time-dependent fitness. The precise bias is likely to depend on environmental factors like population density and time of leaf fall. Our current understanding of the ecology of sex in tree-dwelling aphids is likely to be greatly improved by more extensive ecological and theoretical studies of the problem.

7 Distribution and global warming

Regional distribution

About 5,000 species of aphids have so far been described and most of these come from the temperate regions of the world. In the tropics and subtropics whitefly (Aleyrodidae) and coccids and mealy bugs (Coccoidea) seem to replace the aphids. The respective distributions of all three groups are assumed to reflect their ability to survive in the physical conditions that prevail in the temperate and tropical regions. Alternatively, the distribution of aphids is seen as constrained by cyclical parthenogenesis, which is a very successful way of exploiting the short-lived growth flushes of temperate plants (p. 14), but which cannot be adapted to tropical conditions (p. 44). However, this view depends on the assumption that there are no seasons in tropical forests. This is debatable as most trees in the tropics grow and flower at a specific time of the year, although the growth flushes of the different species may not all occur at the same time. In addition, aphids associated with crops have often retained their pest status when introduced from the temperate regions into tropical and subtropical regions of the world. There are also cyclically parthenogentic species, which occur in the tropics and subtropics, well adapted to the climatic conditions that prevail there. For example, in Australia species of *Schoutedenia* (Greenideinae), *Sensoriaphis* and *Neophyllaphis* (Drepanosiphinae) avoid the rigours of summer by producing eggs in spring. In contrast, *Cervaphis scouteniae* (Greenideinae) in the vicinity of Calcutta, India, achieves its highest rates of increase during the hottest (30°C and above) and driest period of the year. Indeed species from all the aphid subfamilies are found in the tropics and subtropics. So it is likely that tropical species of aphids are better adapted to the temperatures they experience there than their temperate equivalents and *vice versa*. There is no reason to suppose that aphids cannot be well adapted to tropical conditions. It is possible that aphids have adapted to the higher temperatures prevailing in the tropics by having higher lower temperature thresholds for development and higher temperature optima than their equivalents in temperate regions. However, a lot more information on the range of temperatures over which aphids thrive needs to be collected, especially for tropical species. Although continuous parthenogenesis is commoner in species living in the tropics and subtropics than in temperate regions, nevertheless sexual forms have been recorded for 20% of the Indian aphid species. Thus, as aphids are able to survive and thrive in the supposedly 'harsh' environment of the tropics and subtropics, and speciation is unlikely to have been limited by a lack of genetic recombination, it is difficult to understand why aphids have not flourished on the rich tropical and subtropical floras.

Aphids differ from most groups of insects by showing an inverse relationship between the number of aphid species and the number of plant species in different parts of the world. There are more aphid species per 1,000 species of plants in the temperate regions than in the tropics and subtropics. It has been postulated that because of the greater plant diversity in the tropics there are few species of plants there that are sufficiently abundant to sustain short-lived host-specific insects such as aphids.

Most aphids are host specific and have similar sensory and dispersal powers; and it is likely that they do not locate their host plants from a distance. Host finding tends to be a random process. Thus the concept of plant apparency is useful when considering why relatively few plant species have aphids and also the somewhat surprising paucity of aphids in the tropics. The hypothesis is that a plant species that is plentiful for long periods of time, and over a wide area, is apparent (easily found by host-seeking insects) and is more likely to have aphids than less apparent species of plants. As aphids have little control over their direction of flight their host plants have to be abundant if they are to find them (Dixon and others, 1987). The aphid faunas and floras of several countries are sufficiently well known to define the form of the relationship between the number of aphid species per plant species, and the number of plant species per unit area of land (fig. 22). A similar relationship can be derived mathematically from the observations that aphids are mostly host specific and find their host plants by random search (Dixon and others, 1987):

$$S_a / S_p \cong K \exp(-C_{crit} S_p) \qquad (1)$$

where S_a is the number of aphid species, S_p the number of plant species, and K is the average number of aphid species

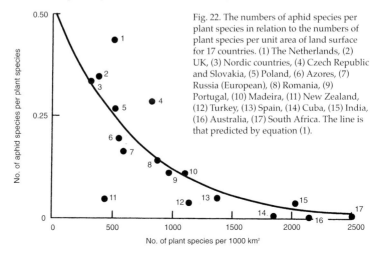

Fig. 22. The numbers of aphid species per plant species in relation to the numbers of plant species per unit area of land surface for 17 countries. (1) The Netherlands, (2) UK, (3) Nordic countries, (4) Czech Republic and Slovakia, (5) Poland, (6) Azores, (7) Russia (European), (8) Romania, (9) Portugal, (10) Madeira, (11) New Zealand, (12) Turkey, (13) Spain, (14) Cuba, (15) India, (16) Australia, (17) South Africa. The line is that predicted by equation (1).

No. of aphid species per plant species

No. of plant species per 1000 km²

per plant species whose cover is equal to or greater than C_{crit} which is the proportional ground cover a plant species must achieve if it is to support an aphid species. Thus in the tropics and subtropics, where the floral diversity is very high (S_p is high), few species of plants are apparent enough to sustain an aphid species (S_a/S_p is low). In the temperate regions there are fewer plant species, and the commoner ones are apparent enough to sustain one or more species of aphid. This may account for the greater number of species of aphids that have evolved in temperate regions. The absence of records of aphids from economically important tropical forest trees such as mahogany and rosewood can hardly be due to negligence by collectors, and suggests that aphids really do not occur on such trees in the tropics. Further advances in our understanding of the distribution of aphids will depend on more information on tropical species and new developments in the analysis of data and theory.

It has been suggested that as aphids originated in the northern hemisphere they have not achieved an equivalent diversity in the southern hemisphere due to the barrier posed by the tropics. This can be tested by determining each country's deviation in number of species of aphids from that predicted by equation (1). The relationship between each country's deficit or excess of aphid species and each country's mean latitude is shown in fig. 23. Although there are aphid fauna lists for very few countries in the southern hemisphere, nevertheless there is a significant correlation indicating a deficit of aphid species in southern latitudes. Thus, the model is a good predictor of aphid species diversity when averaged across countries and strongly suggests that the speciation of aphids has been determined by their limited host location abilities. There are, however, local factors that give rise to variation from this general picture, one of which appears to be latitude, due to a constraint imposed by the area of evolutionary origin. Temperature has a marked effect on the rates of development and growth in aphids (p. 15); they generally develop and grow faster at high than at low temperatures. However, they tend to be adapted to a particular range of temperatures. In the tropics they are exposed to a higher range of temperatures than in the temperate regions, where aphids evolved. There is some evidence to indicate that there is a trade-off between the temperature at which aphids can begin developing and that at which they reach their maximum rate of development and begin to show signs of heat stress. If they can develop at low temperatures, which is advantageous in temperate regions, they cannot thrive at high temperatures and *vice versa*. This may be the basis of the barrier posed by the tropics to aphids colonizing and speciating in that region and providing a source of aphids for the southern latitudes.

Consideration of aphid biology therefore provides an explanation not only for the paucity of aphids in the tropics and subtropics, but also for why so few species of plants are able to sustain aphid species, even in temperate regions.

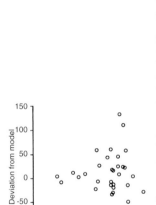

Fig. 23. The effect of latitude of the country on the deviation from the relationship between average area-adjusted species diversity in aphids (S_a) and plants (S_p) predicted by equation (1). (Negative latitude indicates south and positive north.)

Proportionally more host-specific species of aphids are associated with common than with scattered or rare species of plants.

Global warming and distribution

The prediction is that with global warming average temperatures will increase. Thus, if the distribution of aphids is limited, among other things, by the range of temperatures that each species can tolerate then it is likely that the distribution of many aphids in the northern hemisphere will move northwards as temperatures rise. This could occur very rapidly for species for which suitable hosts are already present further north. *Haplocallis pictus*, which characteristically infests oak around the Mediterranean, has recently been recorded in Northern Europe, as have *Aphis solanella* on spindle, *A. spiraecola* on apple, *Brachycaudus divaricatae* on plum and *Dysaphis pyri* on pear (Rakauskas, 2004). But where the host plant distribution has to shift northwards as well, the expectation is that it will take longer. Many southern species of plants are already invading Northern Europe and when abundant will inevitably be colonized by their specific aphids.

Global warming and aphid abundance

Much of the debate about the detrimental effect climate change may have on insect abundance is concerned with its possible differential effect on bud burst in plants and egg hatch in insect herbivores. What is possibly more important, however, is the mechanism by which herbivores and hosts track the seasons and the extent to which herbivores can quickly adapt to changes in the phenology of their host. That insects are well synchronized with their host plants indicates that adaptation is possible. Compared to many insects aphids are a good model group for exploring this problem as the mechanism by which they track the phenology of their host plants and their population dynamics are well understood.

There is a lot of variability in the time of both egg hatch and bud burst, both within and between years. The eggs of the sycamore aphid that hatch at the time of bud burst of sycamore tend to be fitter than those that hatch earlier or later (p. 15). In addition, the timing of egg hatch in aphids appears to be inherited. Spring temperatures affect both bud burst of sycamore and egg hatch in the sycamore aphid. For both trees and eggs there is an inverse relationship between the number of chill days experienced and the thermal time (number of day degrees) (p. 15) from some date in spring to bud burst and egg hatch. Both buds and eggs tend to remain dormant until they experience their chilling requirement, and then the time to bud burst or egg hatch is a function of the intensity of the chilling they have experienced and subsequent spring temperatures. Although the lower developmental thresholds of sycamore and sycamore aphids differ, nevertheless, egg hatch and bud burst are synchronized (p. 15). So although the tracking mechanism used by the plant and insect differ in detail, synchronization is possible. Selection in this case would appear to have resolved a complex optimization problem.

If climate change affects bud burst and egg hatch differentially in this system then the mortality of the aphids hatching in spring is likely to increase and the average fecundity of the survivors to decrease. Climate-change-induced asynchrony in the frequency distributions in time of bud burst and egg hatch in any one year is likely to be slight and any negative effect on aphid numbers will probably be compensated for within the year due to the dynamics of the system. In addition, proportionally more of the survivors will be individuals with inherited responses that enable them to track more closely the phenology of their host plant. These aphids are likely to mate with one another and as a consequence, the following year, proportionally more of the eggs will hatch in synchrony with average bud burst. In other words, selection will quickly correct for any asynchrony between egg hatch and bud burst.

In order to make realistic predictions about the effect of climate change on the abundance of an insect species, it is necessary to take into account the mechanism by which the insect's abundance is regulated and how selection is likely to shape its life-history strategy. For tree-dwelling aphids, where the processes acting at both the individual and the population level are fairly well understood, the evidence indicates that the adverse effect of climate change is unlikely to be as threatening as many people have suggested.

In summary, the greater number of species of aphids in the temperate regions can possibly be accounted for mainly in terms of floral diversity and aphid biology. In the tropics the high floral diversity means relatively few species of plant are common enough to sustain an aphid species. In addition, on the assumption that aphids originated in the northern hemisphere, there is evidence to support the idea that the failure of aphids to achieve an equivalent diversity in the southern hemisphere is due to its distance from the evolutionary origin of aphids and the temperature barrier posed by the tropics. Global warming is likely to result in a northward shift in the distribution of aphid species in the northern hemisphere and a southward shift in the southern hemisphere. Because of its differential effect on bud burst in plants and egg hatch in insect herbivores, global warming is sometimes perceived as a threat to the abundance of tree-dwelling insects, in particular aphids. However, it is likely that in aphids selection will quickly correct for any asynchrony between egg hatch and bud burst.

8 Identification

Introduction to the keys

Diagnostic features

Aphids are small (1–10 mm long) soft-bodied insects, which
can be winged or wingless. The body is segmented, although
the junction between head and thorax in wingless
individuals is hardly visible (fig. 24).

Fig. 24a. General aphid morphology.
Primary rhinaria (olfactory organs)
are present in both larvae and adults;
secondary rhinaria are present only
in adults.

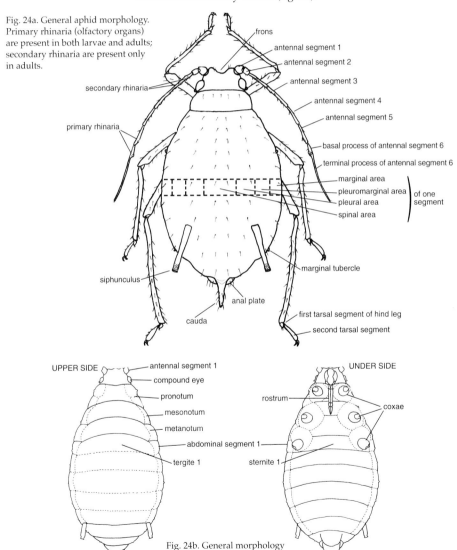

Fig. 24b. General morphology
of wingless adult aphids.

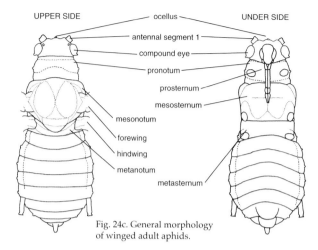

UPPER SIDE   ocellus   UNDER SIDE

antennal segment 1

compound eye

pronotum

prosternum

mesosternum

mesonotum

forewing

hindwing

metanotum

metasternum

Fig. 24c. General morphology
of winged adult aphids.

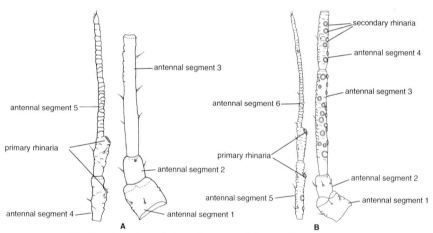

antennal segment 3

antennal segment 5

primary rhinaria

antennal segment 2

antennal segment 4

antennal segment 1

A

secondary rhinaria

antennal segment 4

antennal segment 3

antennal segment 6

primary rhinaria

antennal segment 2

antennal segment 5

antennal segment 1

B

Fig. 24d. General morphology of antennae of adult aphids.
A antenna of unwinged female (fundatrix of *Rhopalosiphum insertum*)
B antenna of winged female (fundatrigenia of *Rhopalosiphum insertum*)

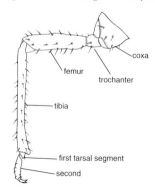

coxa

femur

trochanter

tibia

Fig. 24e. General morphology of
leg of adult aphids.

first tarsal segment

second

The head is bent downward along a sharp edge called the frons. The shape of the frons depends on the frontal tubercles, and is a useful character for identification (fig. 25). The sucking mouthparts consist of four stylets (the mandibles and the maxillae, which together form a tube) enclosed in a four-segmented rostrum, which is placed between and behind the forelegs. The last segment of the rostrum may be subdivided into two parts.

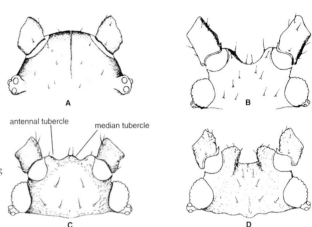

Fig. 25. Shapes of the head.
A antennal and median tubercles not developed, frons round (*Pemphigus bursarius*);
B antennal tubercles well developed, frons diverging (*Monaphis antennata*);
C antennal and median tubercles well developed, frons w-shaped (*Rhopalosiphum padi*);
D antennal tubercles well developed, frons converging (*Myzus cerasi*).

In a few cases (such as the sexual morphs of the Eriosomatini, Fordini and Pemphigini, and males of *Stomaphis*) the rostrum is absent. The antennae have two thick basal segments and a flagellum composed of up to four segments, of which the last is divided into a basal part and usually a thinner terminal process called a 'processus terminalis' (fig. 26).

The thorax is distinctly visible in winged aphids, and contains the muscles that move the wings. In all morphs the

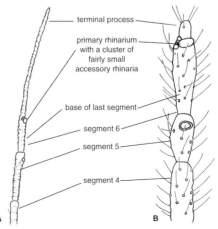

Fig. 26. Antennal segments 4–6.
A with elongate terminal process (*Aphis fabae*)
B with terminal process shorter than base of last antennal segment (*Patchiella reaumuri*)

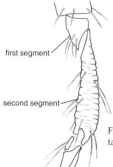

first segment

second segment

thorax bears three pairs of segmented legs. These usually end in two tarsal segments, the relative lengths of which can be a useful character for identification (fig. 27). If wings are present they have a typical venation (fig. 28).

Fig. 27. Tip of hind tibia, with first and second tarsal segments (*Aphis* species)

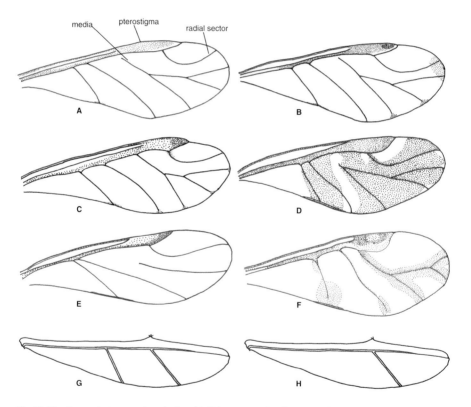

Fig. 28. Venation and pigmentation of wings (A–F fore wings; G–H hind wings)
A with radial sector (R) and two-forked media (M) (*Drepanosiphum acerinum*);
B with radial sector and two-forked media (*Drepanosiphum aceris*);
C with radial sector and two-forked media (*Drepanosiphum dixoni*);
D with radial sector and two-forked media (*Lachnus roboris*);
E with radial sector and unbranched media (*Thecabius affinis*);
F without radial sector and with two-forked media (*Tinocallis platani*);
G with two transverse veins arising separately from longitudinal vein (*Kaltenbachiella pallida*);
H with one transverse vein (*Colopha compressa*).

The abdomen consists of ten segments of which nine are visible. The upper wall of a segment is called a tergite and the lower a sternite. A tail, called the 'cauda', is the original tenth abdominal segment and its shape can also be used for identification (fig. 29). The upper surface of the fifth or sixth abdominal segment bears a pair of pores, which are often placed on tubes called siphunculi, which point

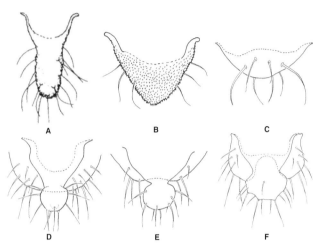

Fig. 29. Shapes of cauda and anal plates from above.
A cauda tongue-like (*Aphis* species);
B cauda triangular (*Dysaphis sorbi*);
C cauda broadly rounded (*Periphyllus testudinaceus*);
D cauda knobbed and anal plate rounded (*Euceraphis punctipennis*);
E cauda knobbed and anal plate notched (*Phyllaphis fagi*);
F cauda knobbed and anal plate two-lobed (*Tinocallis platani*).

upwards and backwards. The shape of the siphunculi and the presence and extent of a net-like sculpture on them can also be used for identification (fig. 30). The anal opening is positioned below the cauda at the junction between the cauda and the last ventral abdominal segment, which is called the anal plate. This is visible in both immature and mature aphids. In front of the anal plate is the genital plate, which is present only in mature aphids and is a useful

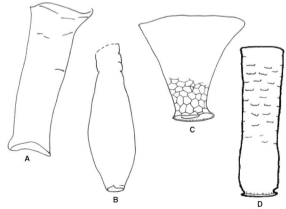

Fig. 30. Shapes of siphunculi (shown with tips pointing downwards)
A cylindrical (*Pterocomma populeum*);
B swollen (*Pterocomma salicis*);
C truncate with polygonal net-like sculpturing below flange (*Periphyllus testudinaceus*);
D slightly swollen and with a constriction below the flange (*Rhopalosiphum padi*).

character for separating larvae and mature aphids (fig. 31). Between the genital and anal plate are 2–4 small hair-bearing processes, called gonapophyses, which are the rudiments of the ninth segment.
    The body may be covered with a coating of powdery

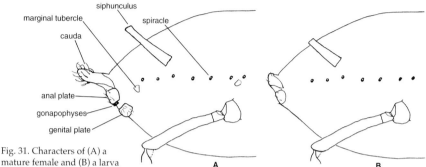

Fig. 31. Characters of (A) a mature female and (B) a larva of *Aphis* species in side view. The small lobes at the front margin of the anal plate, called gonapophyses, and the genital plate occur only in mature females.

wax, or threads of wax, produced by wax glands, which consist of simple tubercles or groups of pores on plates, which are typically distributed segmentally. Depending on the species the abdomen is hardened (sclerotized) and pigmented. Sclerotic and dark pigmented upper areas may be small or absent from some or all tergites. Small sclerotic areas surrounded by thin cuticle are called sclerites, which can be subdivided into spinal, pleural and marginal sclerites.
    An easily visible character of aphids is their colour, which is determined by the following components:

1. Primary colour
2. Pigmentation of the cuticle
3. Waxy coating

In unpigmented aphids or those without a wax covering the primary colour dominates. It reflects the colour of the fluid in the body cavity, and can be black, brown, blue, green, red or nearly colourless (appearing white). Pigmentation may cover the whole of the upper surface, in which case the aphids appear to be black. But in other cases this pigmentation is reduced to patches, which often form a typical pattern. Wax can greatly influence the colour. If the whole body is covered with thick wax then the aphid appears to be white. A covering of waxy powder or a thin layer of wax alters, but does not completely obscure, the primary colour.

Terms

alata (plural alatae): winged individual
anholocyclic: reproducing only by parthenogenesis
aptera (plural apterae): wingless individual
fundatrigenia: second generation after egg hatch, often winged
fundatrix (plural fundatrices): a stem mother (clone founder). It is usually wingless and can be distinguished from the following generations by the greater size of the body and by a tendency towards the shortening of its antennae
gynopara (plural gynoparae): a winged sexupara that gives birth to mating females only
heteroecious: having a primary and a secondary host plant species
holocyclic: reproducing by means of sexuales in autumn and by parthenogenesis for the rest of the year (that is, exhibiting cyclical parthenogenesis)
monoecious: feeding on only one host plant species
morph: one of the different forms present within a species
oviparous: laying eggs
parthenogenesis: production of females by unmated females
sexupara (plural sexuparae): an aphid that gives birth to both males and sexual females
viviparous: giving birth parthenogenetically to live young

Preparing aphids for identification

In order to identify aphids correctly, it is nearly always necessary to study them carefully under a microscope. This is especially true of aphids belonging to difficult or large genera like *Aphis* and *Pemphigus* in which the diagnostic characters are very small. A magnification of up to 500 times is usually sufficient, and phase-contrast lighting is an advantage.

For examination under a microscope the specimens need to be mounted on microscope slides. This must be done carefully. Attempts to save time may result in slides that are unsuitable for future reference or will be less useful when the specimens become distorted. (If you are short of time, have little experience, or are poorly equipped for producing well-mounted specimens and need the help of a specialist, you should send your specimen unmounted. Remounting badly mounted or damaged aphids is time consuming and most specialists will not be prepared to do this.)

In principle there are several techniques for mounting aphids on slides, but the following can be used for fast identification. Aphids should be orientated on a slide so that all the characters that have to be studied for identification are visible. Therefore, first mount one specimen with the ventral (lower) surface uppermost. This is the primary specimen. If you have several specimens, mount one on its side and a third with its dorsal (upper) side uppermost. It is important to spread the antennae, legs and wings of the primary specimen. Care should be taken to arrange this specimen so that the tarsi are visible and the wings are flat. Mounting is easier and the mount is of better quality if the aphids are first stored in 80–90% ethanol (alcohol). This preserves the aphids and expands their wings, legs and all inter-segmental membranes, and they can be transferred directly from ethanol into the mounting medium.

Aphids that you have just collected from trees and wish to identify, however, should first be placed in a watch glass and sorted under a dissecting microscope into small pale species and large dark species. The two groups should be treated differently. Dark or big aphids, or those that become brittle when stored in ethanol, should first be macerated in a solution of sodium hydroxide (5% NaOH) or potassium hydroxide (5% KOH) . Maceration softens the exoskeleton, extends the inter-segmental membranes, dissolves the body contents and makes dark specimens more transparent.

The maceration process requires care and experience. It can be done as follows. First boil the specimens in 95% ethanol for 1 or 2 minutes (care! do this in a water bath). Then place the specimens in the macerating solution at room temperature. Carefully heat the aphids in the solution, ceasing as soon as legs or wings start to stretch or the solution is just about to boil. It is advisable to puncture the specimen with a fine needle between the bases of the hind legs before macerating it. The caustic solution can then enter the body more freely, and the digested contents can be expelled by gently massaging the abdomen. Following maceration, aphids should be washed by replacing the macerating solution with distilled water and gradually adding 50% ethanol. It is important that the body is fully distended with liquid following maceration and washing. Otherwise the animal collapses in on itself and it is then difficult to distinguish the characters (for example, to separate the plates or hairs on the upper and lower surfaces).

Now the specimens can be transferred directly into a water-based mountant, consisting of 12 g clean gum arabic plus 6.5 ml pure glycerol plus 20 g chloral hydrate plus 20 ml distilled water. Successfully mounted slides can be stored for long periods if they are first placed in an oven at 50°C for one week. If an oven is not available the slide must be placed in a horizontal position for a month. After this treatment the cover glass of each slide should be ringed with lacquer or nail varnish. These slides should be stored in special boxes, in dark and dry conditions at room temperature. If stored in this way the slides will last for 20 to 30 years.

A mounted aphid is of little value if it is not labelled. The information on the label should be as indicated in fig. 32. To help you find the specimens on the slides, especially of small and pale aphids, they should be mounted in the centre of the slide. This is easily done with the help of a template. Use a slide that has been marked in the middle on the

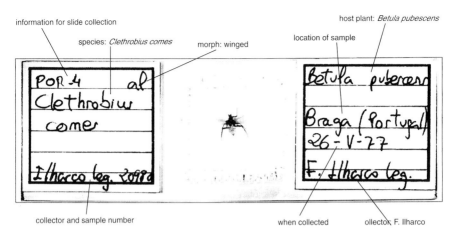

Fig. 32. A labelled slide.

underside, and use this mark to orientate the specimen. The
label should be completed so that the head of the specimen
faces towards you, so that the aphid is correctly orientated
when viewed under a compound microscope.

The keys

Critical identification requires some knowledge of
aphids and access to a reference collection, as species can
vary in colour and in size and shape. This is especially true
of some genera. Two problems must be kept in mind when
using the keys. Because Europe is a climatically diverse
region, both the flora and fauna may vary geographically,
and this is also true of the aphid fauna. Moreover, the keys
and colour photographs do not cover all species, but only
those known to be important (those that commonly occur on
particular deciduous tree species) and those that are
interesting because of their biology. It is therefore essential
that all the characters described in the keys and the species
descriptions (p. 82) are carefully compared with those of the
insect being identified. If you fail to identify your specimen
you may have a species not included in the keys and in this
case you should consult a specialist.

The keys below are for aphids that live on particular
species of deciduous trees, which are arranged in
alphabetical order of the English name. Single aphids can
often be found on non-host plants. Therefore, it is
recommended that initially you collect only aphids that are
present in small colonies or abundant on trees. It is always
better to have several mature specimens from a colony in
order to use the key.

Alder
Alnus glutinosa

antennal segment I

frontal process

II

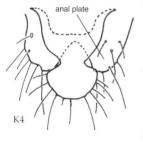

K1

K2

K3

1a A winged female with a balloon-like projection on each side of the prothorax, and a pair of tubercular frontal processes projecting forward between the antennae, as long as or longer than antennal segment 1 (K1). Wingless morphs and immatures with large plate-like frontal and lateral projections, between very short antennae (K2)

Crypturaphis grassii p. 91

1b A winged female without projections on the sides of the prothorax; if frontal processes are present, they are much shorter than antennal segment 1 (K3). If wingless females have frontal projections, then they are not plate-like and the antennae are normal   2

2a Anal plate two-lobed (K4). Populations with wingless females and/or winged females   3

2b Anal plate not lobed, its hind margin rounded (K5). Adults all winged. Hairs on antennal segment 3 much longer than basal diameter of segment (K6)

Clethrobius comes p. 90

anal plate

K4

3a Wingless females with hairs on upper surface of body pigmented; antennal segment 3 with more than 2 long hairs (K7); some hairs on antennal segments 4 and 5 longer than basal diameters of their respective segments

Pterocallis maculata p. 112

anal plate

3b Wingless females with hairs on dorsal surface of body pale; antennal segment 3 with only 1 or 2 long hairs (K8), and all hairs on antennal segments 4 and 5 short and inconspicuous   *Pterocallis alni* p. 112

K5

III

K7

III

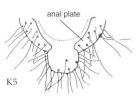

K6

III

K8

Apple
Malus sylvestris

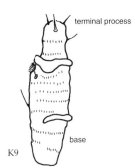

terminal process

base

K9

1a Length of terminal process less than half the length of base of last antennal segment (K9); cauda broadly rounded, usually shorter than its width at base. Wax gland plates with central cells, much smaller than surrounding cells (K10)   *Eriosoma lanigerum* p. 97

1b Length of terminal process more than half the length of base of last antennal segment (usually as long as, or longer than, last antennal segment) (K11); cauda semicircular, helmet-shaped or tongue-shaped, about as long as its basal width, or longer   2

K10

2a Cauda tongue- or finger-shaped, clearly longer than its basal width   3

2b Cauda short, helmet-shaped, semicircular or triangular, not longer than (or about as long as) its basal width when seen from above   5

3a Siphunculi slightly swollen just before the tip and constricted before the well-developed flange at the tip (K12). Marginal tubercles on abdominal segment 7, above and behind the spiracle, and no larger than the spiracular opening (K13)   *Rhopalosiphum insertum* p. 114

3b Siphunculi tapering from base to tip, with flange only moderately developed. Marginal tubercles on abdominal segment 7 below and behind the spiracle, and usually larger than the spiracular opening (K14)   4

terminal process

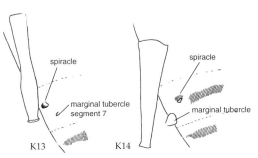

constriction

spiracle

spiracle

marginal tubercle segment 7

marginal tubercle

base

K11   K12   K13   K14

4a Marginal tubercles present on abdominal segments 2–4;
    cauda with 10–19 hairs (rarely fewer than 13) (K15)
                                              Aphis pomi p. 83
4b Marginal tubercles usually absent on abdominal
    segments 2–4; cauda with 7–15 hairs
    (rarely more than 12) (K16)   *Aphis spiraecola* p. 84

5a Antennae of wingless females at least as long as the
    distance from frons to base of siphunculi, and those of
    winged females about as long as the body. Abdominal
    segments in front of siphunculi of wingless females lack
    pigmentation (K17). Galls become yellow to greenish
    yellow and irregularly or transversely curled
                          Dysaphis (Pomaphis) plantaginea p. 95
5b Antennae of wingless females shorter than the distance
    from frons to base of siphunculi, and those of winged
    females shorter than the body. Wingless females with
    pigment spots on the upper surface of the abdomen in
    front of siphunculi (K18). Galls become reddish
    and rolled longitudinally   *Dysaphis devecta* p. 94

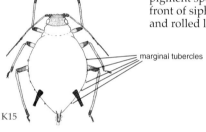

marginal tubercles

K15

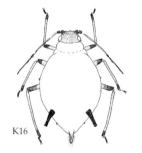

K16

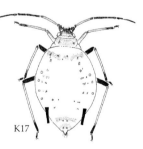

K17

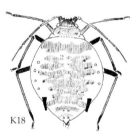

K18

Ash
Fraxinus excelsior

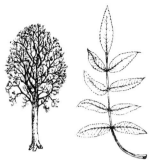

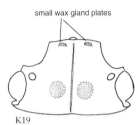

small wax gland plates

K19

Siphunculi absent or only present as small pores. Winged females emerge from leaf-nest galls in spring or early summer. Head usually with a pair of small clear wax gland plates at the front (K19)   *Prociphilus fraxini* p. 111

Beech
Fagus sylvatica

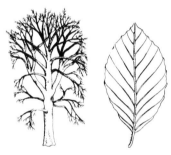

hairy cone

K20

1a Antennae densely hairy, most of the hairs as long as, or longer than, basal diameter of antennal segment 3. Siphunculi on broad dark hairy cones (K20)
Lachnus pallipes p. 100

1b Antennae with sparse, short hairs. Siphunculi pore-like   *Phyllaphis fagi* p. 110

Birch
Betula pendula

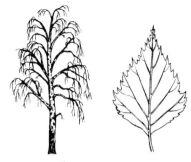

Betula pubescens

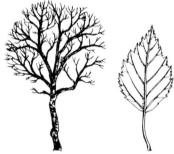

segment 1

segment 2

K21

K22

1a Terminal process equal to or shorter than length of
base of last antennal segment   2

1b Terminal process clearly longer than base of last
antennal segment, and usually more than 1.5 times
as long   5

2a Anal plate with a cleft in the midline dividing it into
two lobes   *Betulaphis quadrituberculata* p. 84

2b Anal plate unlobed, rounded or with a slight
indentation in the midline, but not clearly two-lobed   3

3a Cauda broadly rounded and lacking a terminal knob.
Viviparae both wingless (or with short wings
(brachypterous)) and winged, with 2 or 3 broad white
bands on antennae   *Symydobius oblongus* p. 117

3b Cauda with a terminal knob. Viviparae fully winged
and with un-banded antennae   4

4a Base of last antennal segment usually not more than 1.3
times length of second segment of hind tarsus (K21).
Lives on *Betula pendula*   *Euceraphis betulae* p. 99

4b Base of last antennal segment usually 1.4–1.8 times
as long as second segment of hind tarsus (K22).
Lives on *B. pubescens*   *Euceraphis punctipennis* p. 99

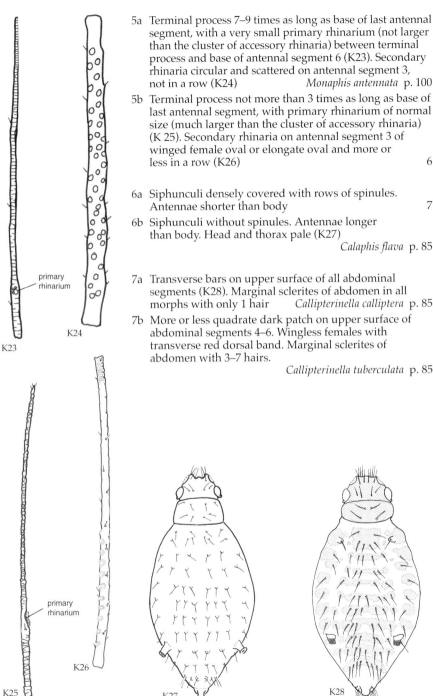

5a Terminal process 7–9 times as long as base of last antennal
    segment, with a very small primary rhinarium (not larger
    than the cluster of accessory rhinaria) between terminal
    process and base of antennal segment 6 (K23). Secondary
    rhinaria circular and scattered on antennal segment 3,
    not in a row (K24)   *Monaphis antennata* p. 100

5b Terminal process not more than 3 times as long as base of
    last antennal segment, with primary rhinarium of normal
    size (much larger than the cluster of accessory rhinaria)
    (K 25). Secondary rhinaria on antennal segment 3 of
    winged female oval or elongate oval and more or
    less in a row (K26)   6

6a Siphunculi densely covered with rows of spinules.
    Antennae shorter than body   7

6b Siphunculi without spinules. Antennae longer
    than body. Head and thorax pale (K27)
                                                    Calaphis flava p. 85

7a Transverse bars on upper surface of all abdominal
    segments (K28). Marginal sclerites of abdomen in all
    morphs with only 1 hair   *Callipterinella calliptera* p. 85

7b More or less quadrate dark patch on upper surface of
    abdominal segments 4–6. Wingless females with
    transverse red dorsal band. Marginal sclerites of
    abdomen with 3–7 hairs.
                                       Callipterinella tuberculata p. 85

primary
rhinarium

K24

K23

primary
rhinarium

K26

K25

K27

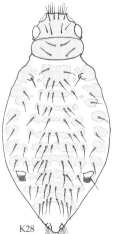

K28

Cherry
Prunus padus (bird cherry)

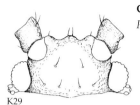

K29

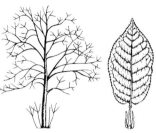

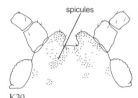

spicules

Prunus avium (wild cherry)

K30

1a Head capsule smooth, wrinkled or rough, but not
ornamented with spicules or nodules (K29)   2

1b Head capsule ornamented with spicules or
nodules (K30)   3

2a Tibiae evenly pigmented. Siphunculi more than 0.3 mm
long (except in fundatrix) and swollen just before the
constriction near the tip (K31). (Fundatrix with terminal
process more than twice as long as base of last
antennal segment)   *Rhopalosiphum nymphaeae* p. 115

2b Tibiae mainly pale, dark only towards the tip. Siphunculi
less than 0.3 mm long or, if longer, without any
discernible swelling near the tip. (Fundatrix with
terminal process not more than twice as long as base
of last antennal segment)   *Rhopalosiphum padi* p. 115

K31

3a Body yellow; siphunculi dark, cylindrical
          Myzus padellus p. 102

3b Body shiny black or dark brown; siphunculi
black, long and tapering   *Myzus cerasi* p. 102

Elder
Sambucus nigra

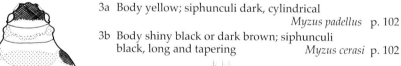

Abdominal segments 1–4 and 7 with large marginal
tubercles, placed between dark marginal plates (K32).
Siphunculi dark, cauda dark and blunt   *Aphis sambuci* p. 83

K32

Elm

Ulmus glabra (English elm)

Ulmus procera (wych elm)

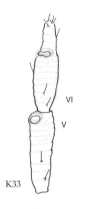

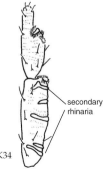

1a Cauda knobbed, anal plate two-lobed. Siphunculi short truncate cones without hairs, dark, with a dark basal plate *Tinocallis platani* p. 119

1b Cauda, if present, broadly rounded, tongue- or finger-shaped. Anal plate, if present, unlobed or with a cleft in the midline but not distinctly two-lobed. Siphunculi, if present, not in form of truncate hairless cones 2

2a Hindwing with 2 transverse veins arising separately from longitudinal vein (fig. 28G, p. 50) 3

2b Hindwing with 1 transverse vein arising from longitudinal vein (fig. 28H, p. 50) 5

3a Siphunculi present as fairly large, often raised pores on abdominal segment 5, with partially hardened rims and surrounded by a ring of hairs. Antennal segment 3 elongate, usually more than 0.8 times head width across eyes 4

3b Siphunculi absent. Antennal segment 3 less than 0.8 times width of head across eyes *Kaltenbachiella pallida* p. 99

4a Antennal segment 5 without secondary rhinaria (K33). (The numbers of secondary rhinaria on the antennal segments are 30–46 on 3rd, 5–9 on 4th, and none on 5th or 6th) *Eriosoma ulmi* p. 98

4b Antennal segment 5 with secondary rhinaria (K34). (Numbers of secondary rhinaria on antennal segments: 18–35 on 3rd, 2–7 on 4th, 1–7 on 5th, and none on 6th) *Eriosoma patchiae* p. 97

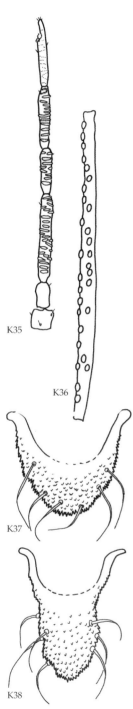

5a Antennal segment 4 similar in length to antennal segment 5 (in 6-segmented antennae; if antenna is 5-segmented then antennal segment 3 more than 3 times as long as antennal segment 4). Media of forewing once-branched   *Colopha compressa* p. 90

5b Antennal segment 4 much shorter than antennal segment 5 (if antenna is 5-segmented then antennal segment 3 less than twice as long as antennal segment 4). Media of forewing unbranched   *Tetraneura ulmi* p. 117

Hawthorn

Crataegus monogyna and *C. oxyacantha*

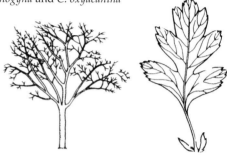

1a Terminal process less than half as long as base of last antennal segment. Antenna of winged female with narrow secondary rhinaria, running across or around the antennae (K35). Siphunculi absent. Wax gland plates with numerous small pores. (Fundatrix with last segments of rostrum shorter than base of antennal segment 5. Spring migrant with pterostigma elongate towards wingtip)
Prociphilus (Stagona) pini p. 111

1b Terminal process more than half as long as base of last antennal segment. Antenna of winged female with oval or circular secondary rhinaria (K36)   2

2a Cauda usually short, helmet-shaped or triangular, shorter than, or about as long as, its basal width seen from above (K37)   3

2b Cauda tongue- or finger-shaped, clearly longer than width at base; sometimes less than 1.4 times its basal width (K38)   6

3a Spring migrant with marginal tubercles on abdominal segment 7 in most specimens, frequently on both sides. Hairs on the upper surface of abdominal segment 8 usually 10–25 μm long   *Dysaphis apiifolia* p. 93

3b Spring migrant without marginal tubercles on abdominal segment 7 in most specimens, and very rarely on both sides. Hairs on the upper surface of abdominal segment 8 usually 30–70 μm long   4

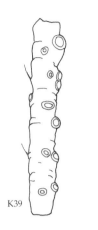

K39

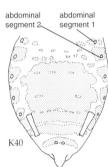

abdominal segment 2   abdominal segment 1

K40

K41

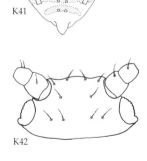

K42

4a Spring migrant with a total of 1–15 secondary rhinaria on antennal segment 5 (K39)   5

4b Spring migrant usually lacking secondary rhinaria on antennal segment 5, or rarely with up to 5. Abdominal segment 2 lacks a dark band   *Dysaphis crataegi* p. 94

5a Abdominal segments 1 and 2 of spring migrant lack dark bands, at most with small scattered circular plates (K40)
                              Dysaphis angelicae p. 93

5b Abdominal segment 2, or both 1 and 2, of spring migrant with a dark cross band, sometimes partly broken into irregular plates (K41)   *Dysaphis ranunculi* p. 94

6a Antennal tubercles absent or weakly developed (K42). Siphunculi of wingless females dark or, if only dark at tip, then short (less than 0.25 mm). Cauda and anal plate dusky to dark. (Terminal process more than twice as long as base of last antennal segment in fundatrix and more than 2.5 times as long in subsequent generations)
                          Rhopalosiphum insertum p. 114

6b Antennal tubercles moderately to well developed. Siphunculi of wingless females pale or dusky, or dark only at tips and then over 0.3 mm long. (Wingless females with inner faces of antennal tubercles parallel or divergent when seen from above)   7

7a Spring migrant with 60–83 secondary rhinaria on antennal segment 3 (K43), 22–57 on 4 and 8–24 on 5   *Ovatus insitus* p. 103

7b Spring migrant with 19–49 secondary rhinaria on antennal segment 3 (K44), 5–20 on 4 and 0–10 on 5
                            Ovatus crataegarius p. 103

K43

K44

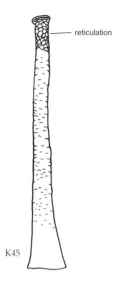

reticulation

K45

Hazel
Corylus avellana

1a Cauda knobbed; anal plate two-lobed. Siphunculi small, truncate cones. Terminal process more than twice, but less than three times, as long as base of last antennal segment. Adult viviparae winged.

Myzocallis coryli p. 102

1b Cauda finger-shaped; anal plate not lobed. Siphunculi long and tubular, with a zone of polygonal net-like patterning (reticulation) near the tip (K45). Terminal process more than 4.3 times as long as base of last antennal segment. Adult viviparae mainly wingless.

Corylobium avellanae p. 90

Hornbeam
Carpinus betulus

Siphunculi short, truncate. Cauda with a knobbed tip; anal plate deeply two-lobed. Adult viviparae all winged

Myzocallis carpini p. 101

Lime
Tilia cordata, T. platyphyllos and *T.* x *vulgaris*

1a Siphunculi absent or present only as inconspicuous pores. (Aphids forming leaf-nest galls)

Patchiella reaumuri p. 105

1b Siphunculi truncate cones, hardly longer than width at base. Cauda knobbed; anal plate two-lobed. (All viviparae winged)   *Eucallipterus tiliae* p. 98

Maple and **sycamore** (*Acer*)

Sycamore    Maple

1a Siphunculi generally stump-shaped, those of wingless
females with a zone of polygonal net-like sculpture near
the tip, and those of winged females generally with more
extensive net-like sculpture. Antennal hairs usually long
and conspicuous, mostly much longer than the diameter
of the segment from which they arise. Winged
females never with dark markings on wings
Periphyllus species 6

1b Siphunculi of variable shape, usually without polygonal
net-like sculpture, or if with some net-like sculpture near
the tip, then adults are all winged with dark markings
on wings. Antennal hairs generally small and
inconspicuous, usually shorter than the segment from
which they arise; if there are long hairs then they are
confined to antennal segment 3 (with occasionally
one also at the base of antennal segment 4)   2

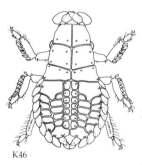

K46

2a Hairs on the front of the head and sides of the abdomen
modified as flattened leaflike plates (K46). (Very small
insects, body length less than 1 mm)
Aestivating nymphs of *Periphyllus* species 6
2b Hairs not modified in this way   3

3a Forewing with a dusky patch at the tip between the ends
of the radial sector and media veins (fig. 28B, p. 50)
Drepanosiphum aceris p. 92
3b Forewing sometimes with dusky spots at the ends of
veins, but these do not extend between vein endings   4

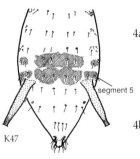

segment 5

K47

4a Abdomen with two broad, dark, cross bands on the
upper surface of segments 4 and 5, that on segment 5
extending at the sides to almost touch the marginal plates
in front of the siphunculi (K47). Pterostigma with a small
dark patch at its outer end. Often adults have short
non-functional wings in summer
Drepanosiphum dixoni p. 92
4b Dark cross bands on abdomen either absent, or shorter,
not extending to marginal plates. Pterostigma lacks a
dark patch. Always fully winged   5

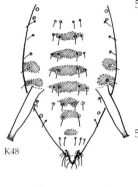

K48

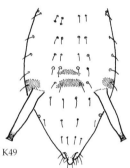

K49

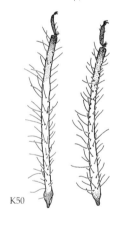

K50

5a Terminal process less than 0.7 times as long as antennal segment 3, 6.0–7.0 times as long as the base of the last antennal segment. Last rostral segment at least 0.14 mm long. Body length 3.1–4.3 mm. With or without a black patch in front of each siphunculus (K48). (Either with no markings on the back or sides of the abdomen (spring/summer), or with a series of brown-black cross bars and side patches including a large patch in front of each siphunculus (autumn))
Drepanosiphum platanoidis p. 92

5b Terminal process more than 0.7 times as long as antennal segment 3, 7.5–12.0 times as long as the base of the last antennal segment. Last rostral segment less than 0.14 mm long. There is always a black patch in front of each siphunculus (K49). (As in the previous species there is marked seasonal variation in the development of dark pigmentation (p. 17))   *Drepanosiphum acerinum* p. 91

6 **Periphyllus on field maple** (*Acer campestre*)

1a Cauda broadly rounded (crescent-shaped), less than half as long as its basal width. Terminal process 2.0–3.6 times as long as the base of the last antennal segment of wingless females (1.9–4.1 in winged females). (Hind tibiae with a pale middle region, contrasting with the dark base and tip (K50))
Periphyllus testudinaceus p. 109

1b Cauda broadly tongue-shaped or with a constriction near the base, more than half as long as its basal width. Terminal process 3.2–6.9 times as long as the base of the last antennal segment of wingless females (3.7–6.7 in winged females).   2

2a The 2 hairs on the base of antennal segment 6 very unequal in length, the longer more than 4 times as long as the shorter, which is less than half the length of the base of antennal segment 6 (including the primary rhinarial complex) (K51)   *Periphyllus hirticornis* p. 109

2b The 2 hairs on the base of antennal segment 6 both long and fine, the longer 1.3–3.0 times the shorter, which is more than half the length of the base of antennal segment 6 (including the primary rhinarial complex) (K52)
Periphyllus obscurus p. 109

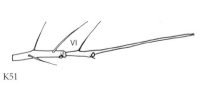

K51

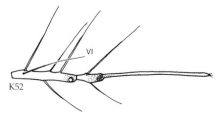

K52

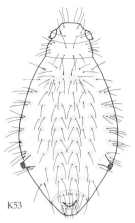

K53

Periphyllus on sycamore (*Acer pseudoplatanus*)

1a The longer of 2 hairs on the base of antennal segment 6 less than half as long as the base of antennal segment 6. (Wingless females with a clear pattern of dark plates on the upper surface of the abdomen (K53). Winged females normally with only 6 long hairs on each of abdominal segments 1–7. Tibiae with a very pale middle region contrasting with the dark base and tip)
Periphyllus testudinaceus p. 109

1b The longer of 2 hairs on the base of antennal segment 6 more than half as long as the base of antennal segment 6. (Siphunculi of wingless females pale, a little longer than their basal diameters. Terminal process less than 3 times as long as the base of the last antennal segment)
Periphyllus acericola p. 108

Oak

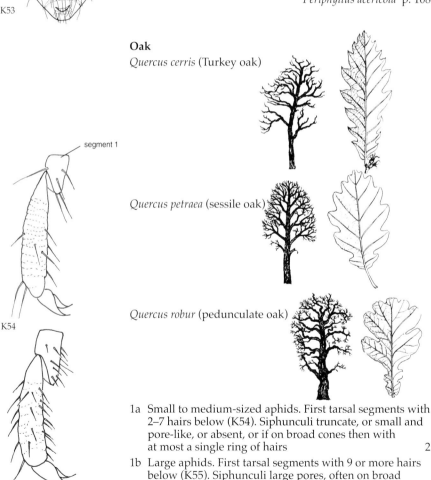

Quercus cerris (Turkey oak)

segment 1

Quercus petraea (sessile oak)

Quercus robur (pedunculate oak)

K54

K55

1a Small to medium-sized aphids. First tarsal segments with 2–7 hairs below (K54). Siphunculi truncate, or small and pore-like, or absent, or if on broad cones then with at most a single ring of hairs 2

1b Large aphids. First tarsal segments with 9 or more hairs below (K55). Siphunculi large pores, often on broad conical bases clothed with numerous hairs 3

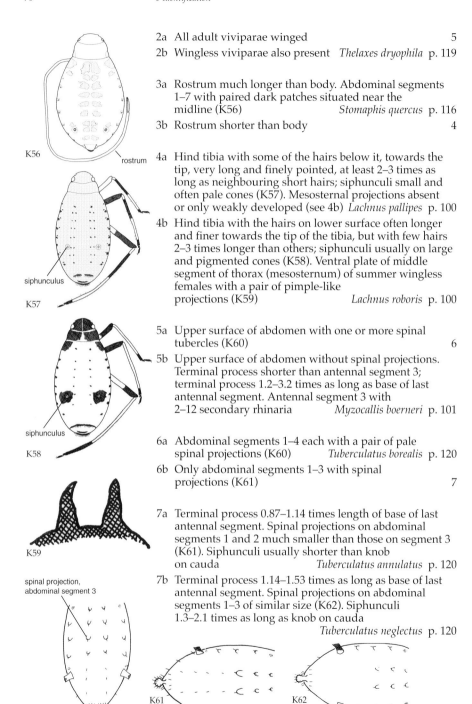

2a All adult viviparae winged   5
2b Wingless viviparae also present   *Thelaxes dryophila* p. 119

3a Rostrum much longer than body. Abdominal segments
    1–7 with paired dark patches situated near the
    midline (K56)   *Stomaphis quercus* p. 116
3b Rostrum shorter than body   4

K56
                                                        rostrum

4a Hind tibia with some of the hairs below it, towards the
    tip, very long and finely pointed, at least 2–3 times as
    long as neighbouring short hairs; siphunculi small and
    often pale cones (K57). Mesosternal projections absent
    or only weakly developed (see 4b) *Lachnus pallipes* p. 100
4b Hind tibia with the hairs on lower surface often longer
    and finer towards the tip of the tibia, but with few hairs
    2–3 times longer than others; siphunculi usually on large
    and pigmented cones (K58). Ventral plate of middle
    segment of thorax (mesosternum) of summer wingless
    females with a pair of pimple-like
    projections (K59)   *Lachnus roboris* p. 100

siphunculus

K57

5a Upper surface of abdomen with one or more spinal
    tubercles (K60)   6
5b Upper surface of abdomen without spinal projections.
    Terminal process shorter than antennal segment 3;
    terminal process 1.2–3.2 times as long as base of last
    antennal segment. Antennal segment 3 with
    2–12 secondary rhinaria   *Myzocallis boerneri* p. 101

siphunculus

K58

6a Abdominal segments 1–4 each with a pair of pale
    spinal projections (K60)   *Tuberculatus borealis* p. 120
6b Only abdominal segments 1–3 with spinal
    projections (K61)   7

7a Terminal process 0.87–1.14 times length of base of last
    antennal segment. Spinal projections on abdominal
    segments 1 and 2 much smaller than those on segment 3
    (K61). Siphunculi usually shorter than knob
    on cauda   *Tuberculatus annulatus* p. 120
7b Terminal process 1.14–1.53 times as long as base of last
    antennal segment. Spinal projections on abdominal
    segments 1–3 of similar size (K62). Siphunculi
    1.3–2.1 times as long as knob on cauda
                                   Tuberculatus neglectus p. 120

K59

spinal projection,
abdominal segment 3

K60

K61

K62

abdominal segment 3   2   1

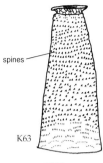

spines

K63

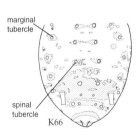

K64

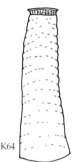

marginal tubercle

K65

marginal plate

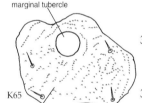

marginal
tubercle

spinal
tubercle

K66

marginal
tubercle

marginal
plate

K67

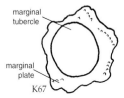

Pear
Pyrus communis

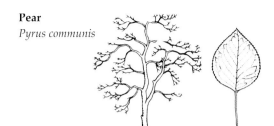

1a Siphunculi with closely-spaced rows of densely-packed little spines (K63)   2

1b Siphunculi black, without densely-packed little spines (K64)   *Dysaphis (Pomaphis) pyri*   p. 96

2a Fundatrix with terminal process longer than base of last antennal segment and with well-developed marginal plates, which are mostly more than twice the diameter of any tubercles present (K65). Winged female with spinal and marginal tubercles usually on abdominal segments 1–7 (K66)   3

2b Fundatrix with terminal process shorter than base of last antennal segment and with marginal tubercles on small plates mostly less than twice the diameter of the tubercles (K67). Winged female with spinal and marginal tubercles on abdominal segments 1–5 (K68)
     Anuraphis farfarae   p. 82

3a Fundatrix lacks spinal and marginal tubercles. Siphunculi of winged female with 18–27 rows of little spines (K69); antennal segment 3 with about 50–60 rhinaria
     Anuraphis catonii   p. 82

3b Fundatrix with large marginal tubercles on abdominal segments 1–7 and often also with spinal tubercles. Siphunculi of winged female with 25–35 rows of little spines (K70); antennal segment 3 with about 70–105 rhinaria   *Anuraphis subterranea*   p. 82

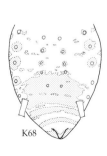

K68

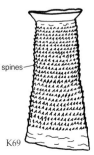

spines

K69   K70

Poplar
Populus alba (white poplar)

Populus nigra (including 'Italica')
(black poplar, including Lombardy poplar)

Populus tremula (aspen)

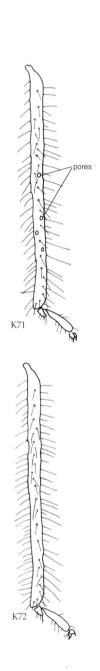

pores

K71

K72

On leaves or stems, not in galls   **Key A**
In galls (only winged females are keyed out)   **Key B**

Key A Aphids on leaves and stems of poplar trees

1a Siphunculi in form of small truncate cones or short
    cylinders (usually at least half as long as basal width),
    with net-like sculpturing at least towards the tip.
    Cauda knobbed or rounded   2
1b Siphunculi without net-like sculpturing; in form of flat
    cones (less than half as long as basal width), or tubular,
    or pore-like, or absent. Cauda rounded   5

2a Hind tibia with 1–22 scattered small pores (K71)
                              Chaitophorus populeti p. 87
2b Hind tibia without pores (K72)   3

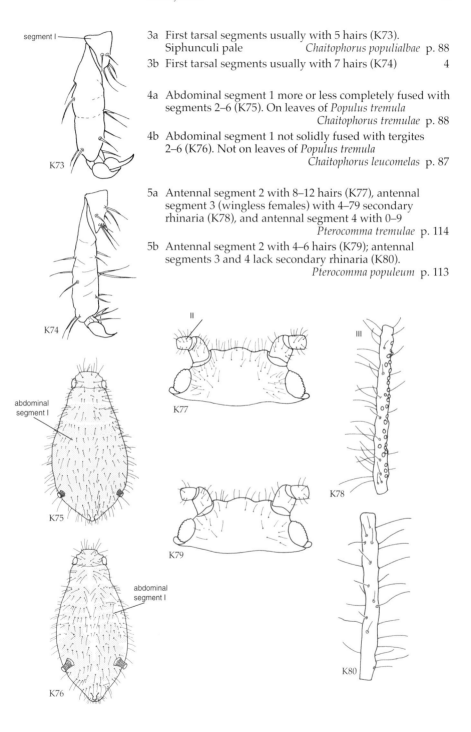

3a First tarsal segments usually with 5 hairs (K73).
    Siphunculi pale   *Chaitophorus populialbae* p. 88
3b First tarsal segments usually with 7 hairs (K74)   4

4a Abdominal segment 1 more or less completely fused with
    segments 2–6 (K75). On leaves of *Populus tremula*
                              Chaitophorus tremulae p. 88
4b Abdominal segment 1 not solidly fused with tergites
    2–6 (K76). Not on leaves of *Populus tremula*
                              Chaitophorus leucomelas p. 87

5a Antennal segment 2 with 8–12 hairs (K77), antennal
    segment 3 (wingless females) with 4–79 secondary
    rhinaria (K78), and antennal segment 4 with 0–9
                              Pterocomma tremulae p. 114
5b Antennal segment 2 with 4–6 hairs (K79); antennal
    segments 3 and 4 lack secondary rhinaria (K80).
                              Pterocomma populeum p. 113

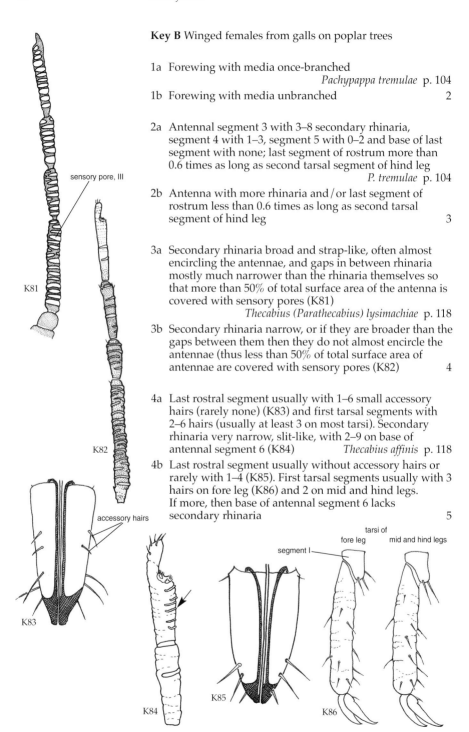

Key B Winged females from galls on poplar trees

1a Forewing with media once-branched
                    Pachypappa tremulae p. 104
1b Forewing with media unbranched   2

2a Antennal segment 3 with 3–8 secondary rhinaria,
    segment 4 with 1–3, segment 5 with 0–2 and base of last
    segment with none; last segment of rostrum more than
    0.6 times as long as second tarsal segment of hind leg
                    P. tremulae p. 104
2b Antenna with more rhinaria and / or last segment of
    rostrum less than 0.6 times as long as second tarsal
    segment of hind leg   3

3a Secondary rhinaria broad and strap-like, often almost
    encircling the antennae, and gaps in between rhinaria
    mostly much narrower than the rhinaria themselves so
    that more than 50% of total surface area of the antenna is
    covered with sensory pores (K81)
                    Thecabius (Parathecabius) lysimachiae p. 118
3b Secondary rhinaria narrow, or if they are broader than the
    gaps between them then they do not almost encircle the
    antennae (thus less than 50% of total surface area of
    antennae are covered with sensory pores (K82)   4

4a Last rostral segment usually with 1–6 small accessory
    hairs (rarely none) (K83) and first tarsal segments with
    2–6 hairs (usually at least 3 on most tarsi). Secondary
    rhinaria very narrow, slit-like, with 2–9 on base of
    antennal segment 6 (K84)   *Thecabius affinis* p. 118
4b Last rostral segment usually without accessory hairs or
    rarely with 1–4 (K85). First tarsal segments usually with 3
    hairs on fore leg (K86) and 2 on mid and hind legs.
    If more, then base of antennal segment 6 lacks
    secondary rhinaria   5

sensory pore, III

K81

K82

accessory hairs

K83

K84

segment I

K85

K86

tarsi of
fore leg   mid and hind legs

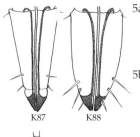

K87　　K88

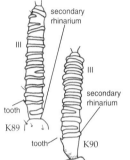

secondary rhinarium

III

III

secondary rhinarium

tooth

K89

tooth

K90

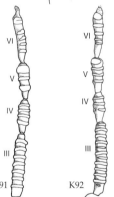

VI

VI

V

V

IV

IV

III

III

K91

K92

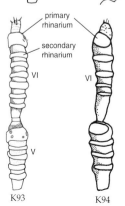

primary rhinarium

secondary rhinarium

VI

VI

V

K93

K94

5a Embryos developing within adult aphids lack mouthparts. Secondary rhinaria always absent from antennal segments 5 and 6. Siphunculi absent. (Sexuparae of this non-host-alternating species develop in the gall)　　*Pemphigus spyrothecae* p. 107

5b Embryos developing within adult aphids with mouthparts. Secondary rhinaria usually present on antennal segments 5 and 6. Siphunculi present as small pores　　6

6a Last rostral segment 0.10–0.12 mm long, its basal part straight-sided (K87), usually more than 0.55 times as long as 2nd segment of hind tarsus　　7

6b Last rostral segment 0.07–0.10 mm long, its basal part with sides slightly convex (K88), usually less than 0.55 times as long as 2nd segment of hind tarsus　　8

7a Secondary rhinaria narrow, slit-like, extending almost to base of antennal segment 3, the rhinarium nearest the base of antennal segment 3 usually nearer the base than the small tooth on the inner side of the segment (K89); 12–18 secondary rhinaria on segment 3, 3–4 on segment 4, 4–7 on segment 5 and 4–7 on base of segment 6
　　Pemphigus populinigrae p. 106

7b Tooth on antennal segment 3 nearer the base than the basal most secondary rhinarium (K90); 10–14 secondary rhinaria on segment 3, 2–5 on segment 4, 2–4 on segment 5, and 2–8 on base of segment 6
　　Pemphigus protospirae p. 107

8a Antennal segment 3 shorter than antennal segments 4+5 in most specimens (K91); 7–12 secondary rhinaria on antennal segment 3, 3–6 on segment 4, 1–4 on segment 5, and 1–7 on base of segment 6
　　Pemphigus gairi p. 106

7b Antennal segment 3 usually about the same length as, or longer than, antennal segments 4+5 (K92)　　9

9a Primary rhinarium on antennal segment 6 much larger than adjacent secondary rhinaria and forming a broad band that extends more than halfway around antenna (K93). Secondary rhinaria: 12–16 on segment 3, 3–5 on segment 4, 4–6 on segment 5, and 6–8 on base of segment 6　　*Pemphigus passeki* p. 106

9b Primary rhinarium on antennal segment 6 normal or irregular in shape but not extending more than halfway around antenna (K94). Secondary rhinaria: 8–17 on antennal segment 3, 2–8 on segment 4, 2–6 on segment 5, and 3–8 on base of segment 6
　　Pemphigus bursarius p. 105

Rowan and **whitebeam**
Sorbus aucuparia (rowan)

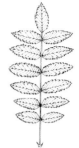

Sorbus aria (whitebeam)

1a Cauda short, helmet-shaped, triangular or semicircular,
    shorter than, or about as long as, width at base seen
    from above   2
1b Cauda tongue- or finger-shaped, clearly longer than
    width at base seen from above
                                          Rhopalosiphum insertum p. 114

2a Siphunculi pale. Marginal tubercles present on
    abdominal segments 1–7 (sometimes absent from 6)
    (K95). Cauda with 6–12 hairs
                                      Dysaphis (Pomaphis) sorbi p. 96
2b Siphunculi dark, at least towards the tip. Marginal
    tubercles present on abdominal segments 1–4 or 1–5,
    absent from 6 and 7 (K96). Cauda with 5, or
    rarely 6, hairs   *Dysaphis (Pomaphis) aucupariae* p. 95

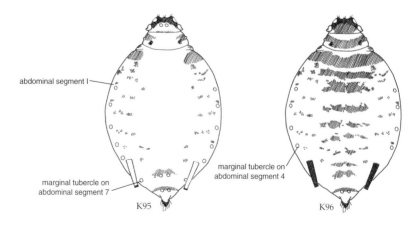

abdominal segment I

marginal tubercle on
abdominal segment 7

marginal tubercle on
abdominal segment 4

K95

K96

PLATE 1

1. Winged adult of *Anuraphis subterranea* on leaf of pear in May
2. Fundatrix of *Aphis fabae* with nymphs on leaf of spindle
3. Wingless adult of *Aphis fabae* being attacked by a larva of the gall midge, *Aphidoletes* species
4. Colony of *Aphis fabae evonymi*, a brown subspecies of *A. fabae*, on spindle
5. Colony of *Aphis farinosa* on willow, with a parasitic wasp (top centre)
6. Colony of *Aphis pomi* on apple in June

PLATE 2

1. Immature oviparous female of *Aphis pomi* on apple
2. Colony of *Aphis sambuci* on elder
3. Wingless adult of *Aphis spiraecola* with nymphs
4. Nymph of *Betulaphis quadrituberculata* on the underside of a birch leaf
5. Wingless adult of *Callipterinella tuberculata* giving birth to a nymph
6. Oviparous female of *Callipterinella tuberculata* laying an egg on birch

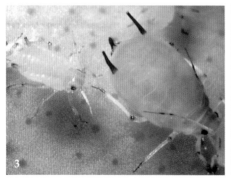

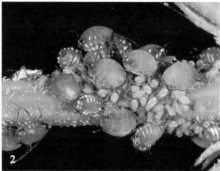

PLATE 3

1. Colony of *Callipterinella tuberculata*, with attendant ant, on a birch twig
2. Egg of *Cavariella aegopodii* inserted between a bud and a willow twig
3. Colony of *Cavariella aegopodii* on a willow leaf, with a hatched syrphid egg in the bottom right hand corner
4. Colony of *Cavariella pastinacae* in summer on the underside of a willow leaf
5. Wingless adult of *Cavariella pastinacae* in summer on the underside of a willow leaf
6. Colony of *Chaitophorus leucomelas* on a poplar leaf

PLATE 4

1. Wingless adults of *Chaitophorus populeti* on a poplar leaf
2. Colony of *Chaitophorus populeti* on a young stem of *Populus tremula*
3. Colony of *Chaitophorus populeti* on a young stem and leaves of *Populus tremula*
4. Colony of *Chaitophorus tremulae* on the underside of a leaf of *Populus tremula*
5. Wingless adult of *Chaitophorus populialbae* with a nymph
6. Nymph of *Chromaphis juglandicola* on the underside of a walnut leaf

 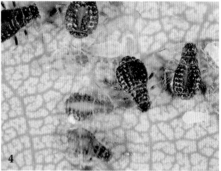

 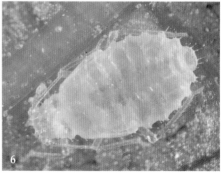

PLATE 5

1. Winged adult of *Chromaphis juglandicola* on the underside of a walnut leaf
2. Gall on the upper surface of an elm leaf induced by *Colopha compressa*
3. Opening on the underside of an elm leaf of a gall induced by *Colopha compressa*
4. Colony of *Crypturaphis grassii* on the underside of an alder leaf
5. Oviparous female of *Crypturaphis grassii* on the underside of an alder leaf
6. Nymph of *Drepanosiphum aceris* on the underside of a field maple leaf

PLATE 6

1. Adult of *Drepanosiphum aceris* on the underside of a field maple leaf
2. Adult of *Drepanosiphum platanoidis* on the underside of a sycamore leaf
3. Colony of *Dysaphis apiifolia petroselini* in a galled hawthorn leaf
4. Winged second generation individual of *Dysaphis apiifolia petroselini* on a hawthorn leaf
5. Red galls induced by *Dysaphis devecta* on the margins of apple leaves
6. Colony of *Dysaphis devecta* inside a gall

PLATE 7

1. Longitudinal leaf roll gall induced on apple by *Dysaphis plantaginea*
2. Colony of *Dysaphis plantaginea* inside a gall
3. Adult of *Dysaphis pyri* with nymphs on the underside of a pear leaf
4. Colony of *Eriosoma lanigerum* on an apple twig
5. Swellings on branches of apple induced by *Eriosoma lanigerum*
6. Galled elm leaf induced by *Eriosoma ulmi*

PLATE 8

1. Nymph of *Eucallipterus tiliae* on the underside of a lime leaf
2. Adult of *Eucallipterus tiliae* on the underside of a lime leaf
3. Oviparous female of *Eucallipterus tiliae*
4. Adult of *Euceraphis betulae* on the underside of a birch leaf
5. Adult of *Euceraphis punctipennis* on a birch leaf
6. Oviparous female of *Euceraphis punctipennis* on a birch leaf

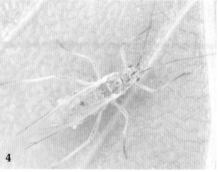

PLATE 9

1. Galls on elm leaves induced by *Kaltenbachiella pallida*
2. Adult of *Lachnus roboris*
3. Oviparous females of *Lachnus roboris* laying eggs on an oak branch
4. Nymph of *Monaphis antennata* on the upperside of a birch leaf
5. Nymph of *Myzocallis boerneri* on the underside of an oak leaf
6. Adult of *Myzocallis boerneri* on the underside of an oak leaf

PLATE 10

1. Winged adults of *Myzocallis coryli* on the underside of a hazel leaf
2. Close up of winged adults and nymphs of *Myzocallis coryli*
3. Leaf nest gall induced on cherry by *Myzus cerasi*
4. Wingless adult of *Myzus cerasi* with a nymph
5. Adult of *Panaphis juglandis* with nymphs on the upper surface of a walnut leaf
6. Rolled leaf galls on lime induced by *Patchiella reaumuri*

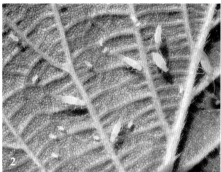

PLATE 11

1. Fundatrix of *Patchiella reaumuri* with nymphs on the petiole of a lime leaf
2. Winged adult of *Patchiella reaumuri* on a lime leaf
3. Gall on the petiole of a poplar leaf induced by *Pemphigus bursarius*
4. Colony of *Pemphigus bursarius* inside a petiole gall (cut open)
5. Gall on the upper surface of the midrib of a poplar leaf induced by *Pemphigus passeki*
6. Colony of *Pemphigus passeki* inside a leaf gall (cut open)

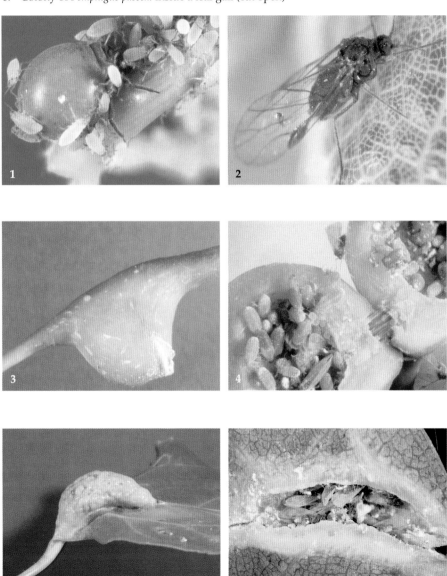

PLATE 12

1. Gall on the upper surface of a poplar leaf induced by *Pemphigus populinigrae*
2. Gall on the petiole of a poplar leaf induced by *Pemphigus protospirae*
3. Gall on the petiole of a poplar leaf induced by *Pemphigus spyrothecae*
4. Group of aestivating nymphs of *Periphyllus acericola* on the underside of a sycamore leaf
5. Group of aestivating nymphs of *Periphyllus aceris* on the under surface of a Norway maple leaf
6. Nymphs of *Periphyllus testudinaceus* on a maple twig

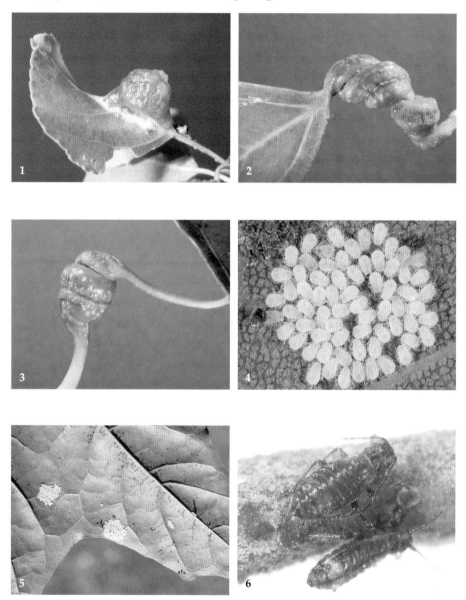

PLATE 13

1. Wax-covered adult and nymph of *Phyllaphis fagi* on the underside of a beech leaf
2. Wax-covered nymphs of *Phyllaphis fagi* on the underside of a beech leaf
3. Leaf nest induced by *Prociphilus fraxini* on ash in May
4. Winged adult of *Prociphilus fraxini* about to leave ash in May
5. Wingless adult and nymphs of *Pterocomma pilosum* on a willow twig
6. Wingless adults of *Pterocomma salicis* on a willow stem

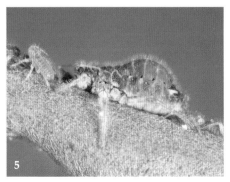

PLATE 14

1. Curled leaf galls induced on apple by *Rhopalosiphum insertum*
2. Nymph of *Rhopalosiphum insertum* in a leaf gall on apple
3. Wingless adult and nymphs of *Rhopalosiphum nymphaeae* on a leaf of water lily, *Nymphaea alba*
4. Fundatrix of *Rhopalosiphum padi* in a leaf gall on bird cherry
5. Winged adult of *Stomaphis longirostris* on the trunk of a willow tree
6. Oviparous female and arostrate male of *Stomaphis longirostris*

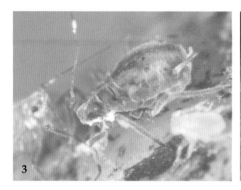

PLATE 15

1. Wingless adults and nymphs of *Stomaphis quercus* in crevices in the bark of oak
2. Colony of *Stomaphis quercus* in a crevice in the bark of oak, attended by ants
3. Galls induced by *Tetraneura ulmi* beginning to develop on the upper surface of an elm leaf
4. Elm leaf with many galls on its upper surface induced by *Tetraneura ulmi*
5. Red gall induced on a poplar leaf by a fundatrix of *Thecabius affinis*
6. Galls induced on poplar leaves by second generation *Thecabius affinis*

PLATE 16

1. Fundatrices and nymphs of *Thelaxes dryophila* on young shoots and a leaf petiole of oak
2. Winged adults of *Tinocallis platani* on the underside of an elm leaf
3. Nymph of *Tinocallis platani* on the underside of an elm leaf
4. Winged adult (dark colour form) of *Tuberculatus annulatus* on the underside of an oak leaf
5. Wingless adult and nymphs of *Tuberolachnus salignus* on a willow stem
6. Winged adult of *Tuberolachnus salignus* on willow

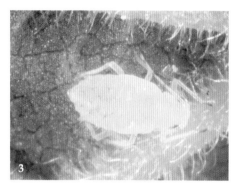

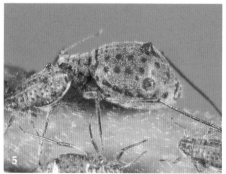

Spindle
Euonymus europaea

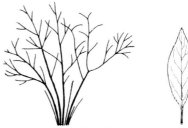

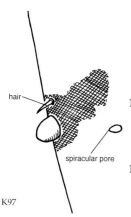

1a Antennal segment 3 of wingless females 11–20 times as long as the longest hair on this segment. Hairs on marginal plates of abdominal segment 3 usually less than twice maximum diameter of the spiracular pore on this segment (K97)   *Aphis solanella* p. 84

1b Antennal segment 3 of wingless females 4–9 times as long as the longest hair on this segment. Hairs on marginal plates of abdominal segment 3 more than twice maximum diameter of the spiracular pore on this segment (K98)   *Aphis fabae* p. 83

Sweet chestnut
Castanea sativa

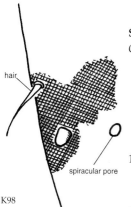

1a Terminal process equal to, or longer than, base of last antennal segment. Cauda knobbed and anal plate divided by a median cleft into two lobes
Myzocallis (Agrioaphis) castanicola p. 101

1b Terminal process not more than half as long as base of last antennal segment. Cauda broadly rounded and anal plate not two-lobed. (Winged female with clear area at base of forewing)   *Lachnus roboris* p. 100

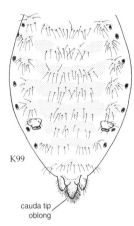

Walnut
Juglans regia

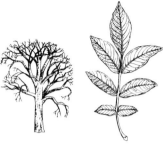

K99

cauda tip
oblong

1a Abdomen with rather broad dark bands across most
segments (K99). Antennal hairs very long; siphunculi
with flange. Cauda knobbed, tip region oblong, with
about 30–45 hairs. Antennal segment 3 more than
4 times as long as antennal segment 6
                                    Panaphis juglandis p. 105

1b Abdomen without broad dark bands across most
segments (K100). Antennal hairs short; siphunculi
without flange. Cauda knobbed, tip transverse, with
about 15–20 hairs. Antennal segment 3 about 2.3–3.1
times as long as antennal segment 6
                                    Chromaphis juglandicola p. 89

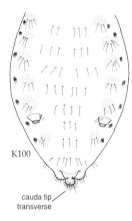

Willow
Salix alba (white willow)

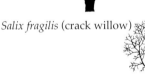

K100

cauda tip
transverse

Salix fragilis (crack willow)

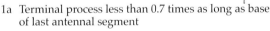
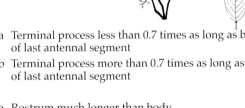

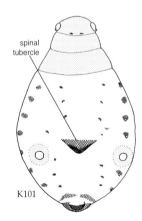

spinal
tubercle

1a Terminal process less than 0.7 times as long as base
of last antennal segment   2

1b Terminal process more than 0.7 times as long as base
of last antennal segment   3

2a Rostrum much longer than body
                                    Stomaphis longirostris p. 116

2b Rostrum much shorter than body. Abdomen with a
large dark spinal tubercle (K101)
                                    Tuberolachnus salignus p. 121

K101

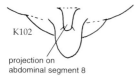

K102

projection on
abdominal segment 8

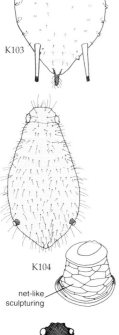

K103

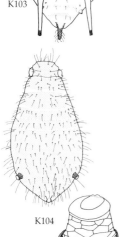

K104

net-like
sculpturing

K105

3b Body hairs sparse and usually rather short, at least on the upper surface of the body 4

3a Body usually densely clothed with hairs, which are either long and fine, or thick; if hairs are short and sparse in the midline of the upper surface then the antennae, legs and cauda are densely hairy 5

4a Abdominal segment 8 with a backwardly-directed projection above the cauda bearing a pair of hairs near the tip (K102) 6

4b Abdominal segment 8 without a projection above the cauda (K103) *Aphis farinosa* p. 83

5a Small to medium-sized. Siphunculi stump-shaped, usually with net-like sculpturing (K104). Cauda knobbed, or rounded, or bluntly triangular, with 6–14 hairs 9

5b Medium to large (K105). Siphunculi almost cylindrical or swollen, usually without net-like sculpturing. Cauda rounded or bluntly triangular with about 20–60 hairs, or tongue-shaped with a constriction and fewer hairs 11

6a Siphunculi cylindrical or tapering, not swollen (K106) *Cavariella theobaldi* p. 87

6b Siphunculi club-shaped or at least slightly swollen (K107) 7

7a Terminal process more than 2.5 times as long as base of last antennal segment. Siphunculi long, club-shaped, smooth or weakly imbricated (that is, with a tile-like pattern of sculpture), projection on abdominal segment 8 broadly tongue-shaped, frequently with flat tip (K107) *Cavariella pastinacae* p. 86

7b Terminal process less than 2.4 as long as base of last antennal segment. Siphunculi various 8

K106

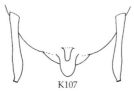

K107

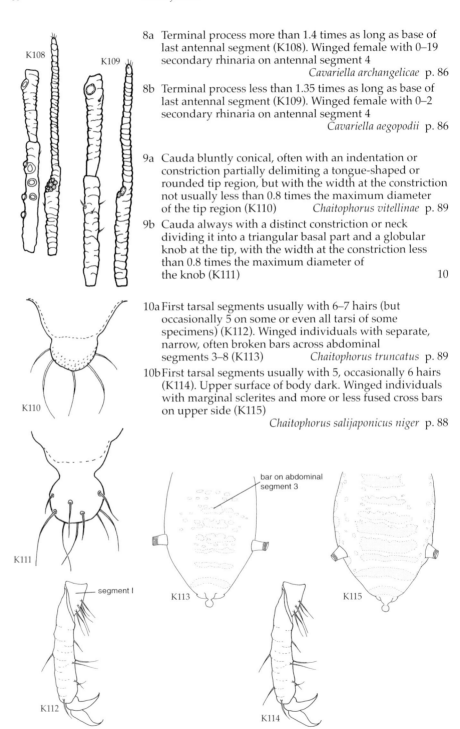

8a   Terminal process more than 1.4 times as long as base of last antennal segment (K108). Winged female with 0–19 secondary rhinaria on antennal segment 4
Cavariella archangelicae p. 86

8b   Terminal process less than 1.35 times as long as base of last antennal segment (K109). Winged female with 0–2 secondary rhinaria on antennal segment 4
Cavariella aegopodii p. 86

9a   Cauda bluntly conical, often with an indentation or constriction partially delimiting a tongue-shaped or rounded tip region, but with the width at the constriction not usually less than 0.8 times the maximum diameter of the tip region (K110)   *Chaitophorus vitellinae* p. 89

9b   Cauda always with a distinct constriction or neck dividing it into a triangular basal part and a globular knob at the tip, with the width at the constriction less than 0.8 times the maximum diameter of the knob (K111)   10

10a First tarsal segments usually with 6–7 hairs (but occasionally 5 on some or even all tarsi of some specimens) (K112). Winged individuals with separate, narrow, often broken bars across abdominal segments 3–8 (K113)   *Chaitophorus truncatus* p. 89

10b First tarsal segments usually with 5, occasionally 6 hairs (K114). Upper surface of body dark. Winged individuals with marginal sclerites and more or less fused cross bars on upper side (K115)
Chaitophorus salijaponicus niger p. 88

bar on abdominal segment 3

segment I

11a Siphunculi up to half as long as 2nd segment of hind
    tarsus, rounded at tip and lacking a flange (K116)
                                        Plocamaphis amerinae p. 110
11b Siphunculi truncate at the tip, usually with at least a
    small flange (K117)   12

12a Antennal segment 2 with 8–20 hairs. Antennal segment 3
    with a few (1–13) secondary rhinaria. Siphunculi
    slightly to markedly swollen, red (K117)
                                        Pterocomma salicis p. 114
12b Antennal segment 2 with 3–6 hairs. Antennal segment 3
    never with secondary rhinaria. Siphunculi swollen (K118)
    or cylindrical, yellowish   13

13a Base of antennal segment 6 with 5–10 hairs longer than
    diameter of the segment, plus 2–4 short hairs (K120).
    Siphunculi cylindrical (K119)   *Pterocomma pilosum* p. 113
13b Base of antennal segment 6 with only 1–3 long hairs
    (K121), plus 2–4 short hairs. Siphunculi cylindrical
    or swollen   *Pterocomma rufipes* p. 113

K116

K117

K118

K119

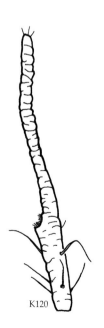

K120

K121

Descriptions of genera and species

Genus *Anuraphis*

There are nine species in this genus. At least some of them host alternate, with pear (*Pyrus*) as the primary host. On the primary host the fundatrices and their offspring feed on the undersides of leaves, which bend down along the midrib so that the two halves touch each other, forming a pseudo-gall. The aphids are brown or green, with black, truncate siphunculi, which have close-set rows of fine spinules across them and well developed flanges.

Anuraphis catonii Hille Ris Lambers

A host-alternating species, with pear (*Pyrus*) as the primary host and *Pimpinella* species as the secondary host. On pear the fundatrices and their offspring are greenish brown. On the secondary host this aphid is attended by ants.

Anuraphis farfarae (Koch)

A host-alternating species, with pear (*Pyrus*) as the primary host and coltsfoot (*Tussilago*) as the secondary host. On pear the fundatrices are brown and their offspring green. On the secondary host this aphid is attended by ants.

Anuraphis subterranea (Walker) **Pear–hogweed aphid (Pl. 1.1)**

A host-alternating species, with pear (*Pyrus*) as the primary host and Apiaceae (parsnip, *Pastinaca* and hogweed, *Heracleum*) as secondary hosts. On pear the fundatrices and their offspring are brown and on the secondary host the colonies are attended by ants.

Genus *Aphis*

This is a large genus, with more than 400 species in the world. The morphological differences between species are small, but the species differ greatly in their host plants and overwintering strategies. Some groups of closely related species are associated with particular families of plants, whereas others live on distantly related plants. Some species are host-alternating, others not. A number of species are able to reproduce parthenogenetically year after year (they are anholocyclic), whereas most are holocyclic. Most species live on one plant species (they are monophagous), like *A. pomi*, or on one plant genus. Others are oligophagous, like *A. sambuci*, and a few species can colonize plants belonging to several different families (they are polyphagous), like *A. fabae*. Most aphids are small to medium-sized. They are often green, but

can vary in colour from pale to dark olive green, or
sometimes black, brown, red or yellow.

Aphis fabae Scopoli **Black bean aphid (Pl. 1.2, 1.3, 1.4)**

A host-alternating species, with spindle (*Euonymus*), guelder
rose (*Viburnum opulus*) or mock orange (*Philadelphus
coronarius*) as its primary hosts, and many secondary hosts.
On the primary hosts the fundatrices and their offspring are
a dark olive green, with black siphunculi. If not exposed to
sunlight they look black. The fundatrices are fatter than the
wingless viviparous females of the following generations.
Adults have bands on abdominal segments 7 and 8, and
occasionally on more anterior segments, and nymphs have
white waxy stripes across the body. On the primary hosts
this species induces curling of the leaves and the colonies are
attended by ants.

A. fabae is a complex of subspecies, each of which can
colonize a wide range of secondary host plants. One of these
is preferred and is the 'marker host', which can be used to
identify the subspecies.

Aphis farinosa Gmelin **(P1. 1.5)**

This is a non-host-alternating species. It occurs in dense
colonies on the young shoots of willow in spring and early
summer. Wingless viviparous females are dull green, not
powdered with wax, and have pale siphunculi and a
distinctly darker cauda. The colonies are attended by ants.

Aphis pomi de Geer **Green apple aphid (Pl. 1.6, 2.1)**

This is a non-host-alternating holocyclic species, which
induces leaf curls on various genera of woody Maloideae.
Wingless viviparous females are green, not powdered with
wax, and have black siphunculi and cauda. Later generations
colonize the undersides of leaves and are yellowish green.
The colonies are attended by ants.

Aphis sambuci Linnaeus **Elder aphid (Pl. 2.2)**

This species has dark green or brown wingless females,
which live in dense colonies near the tips of young green
twigs of elder (*Sambucus* species), its primary host. It host
alternates and spends summer on the roots and root collars
of various plants. On the primary host the fundatrices have a
slight waxy powdering. Wingless viviparous females have
black antennae, legs, siphunculi and cauda. Adults have
cross bands on abdominal segments 7 and 8. Adults and
immature aphids frequently have white waxy stripes across

the sides of their abdominal segments. The colonies are frequently attended by ants.

Aphis solanella Theobald

A host-alternating species, with spindle (*Euonymus*), or occasionally mock orange (*Philadelphus coronarius*), as its primary host, and many secondary hosts. On the primary host the colour of the fundatrices and their offspring is similar to that of *A. fabae*. The colonies are attended by ants.

Aphis spiraecola Patch **Green citrus aphid (Pl. 2.3)**

The green or yellow-green wingless females of this host-alternating species live in dense colonies, often curling and distorting flower heads and the leaves at the tips of stems. It lives on *Spiraea* species, its primary host, and a wide variety of secondary host plants, including apple. Aphid infestation of young plants often results in reduced growth. This species is morphologically very similar to *A. pomi*, but differs in that its last rostral segments are shorter, and there are fewer hairs on the cauda and fewer marginal tubercles. *A. spiraecola* reproduces better than *A. pomi* at high temperatures and is less susceptible to many insecticides. Colonies of this species are attended by ants.

Genus *Betulaphis*

There are fewer than ten species in this genus. They feed on birch and are usually not attended by ants. They are small, rather flat and oval, with short six-segmented antennae, and the terminal process is about as long as the base of the last antennal segment. Viviparous females have a short, conical cauda and a two-lobed anal plate.

Betulaphis quadrituberculata (Kaltenbach) **(Pl. 2.4)**

This species is non-host-alternating and holocyclic, and lives on the undersides of leaves of various species of birch, especially *Betula pubescens*. Wingless viviparous females are whitish yellow or pale yellowish green, with the tarsi or the tips of the tarsi dark. Winged females sometimes have a black spot on the upper surface.

Genus *Calaphis*

This is a genus with 13 species. Most feed on birch and are usually not attended by ants. They are fragile, with long thin legs and well developed antennal tubercles. The antennae are longer than the body and the terminal process is longer than the base of the last antennal segment. Viviparous females have a slightly constricted cauda and a two-lobed anal plate.

Calaphis flava Mordvilko

This species does not host alternate. It is holocyclic and lives on various species of birch, especially *Betula pubescens*. Young leaves are preferred and *C. flava* also colonizes the young growth of other species of birch. Wingless viviparous females are pale green or yellowish, with dark tips to the antennal segments and tibiae, and dark tarsi. The siphunculi are pale, rarely with dark tips. Winged females have dark wing veins.

Genus *Callipterinella*

This genus includes three species, all of which live on birch. They are usually attended by ants. The body is covered above with long, strong, pointed or blunt hairs. The antennae are shorter than the body and the terminal process is longer than the base of the last antennal segment. Viviparous females have a constricted, knobbed cauda, and short, dark, truncate siphunculi with rows of minute spinules.

Callipterinella calliptera (Hartig)

This non-host-alternating holocyclic species often lives in groups on the undersides of leaves of birch (*Betula verrucosa* and *B. pubescens*) spun together by lepidopterous larvae. Wingless viviparous females are green with dark brown bands across the abdomen. The marginal plates of the abdomen in all morphs bear 3–7 hairs. In the winged females the cross bands are less well developed.

Callipterinella tuberculata (von Heyden) **(Pl. 2.5, 2.6, 3.1)**

This non host-alternating, holocyclic species lives on the undersides of leaves of birch (*Betula verrucosa*). Wingless viviparous females are yellow with a brown head, the front of the abdomen is red, and the rear part of abdomen has a large black spot on its upper side. The marginal plates of the abdomen in all morphs have only one hair. Winged females have small, brown spots on the upper side of the abdomen but no cross bands or larger spots.

Genus *Cavariella*

There are about 30 species in this genus. Most host alternate, with willow as their primary host and various Apiaceae as secondary hosts. The genus is characterized by the presence of two hairs positioned close together on a prominent tubercle, the supracaudal process, which in winged forms is reduced to a rather small wart. The antennae are shorter than the body and the terminal process can be longer or shorter than the base of the last antennal segment. Viviparous females have a tongue-shaped cauda and cylindrical or swollen siphunculi.

Cavariella aegopodii (Scopoli) **Willow–carrot aphid (Pl. 3.2, 3.3)**

This is a host-alternating species. Its primary host is willow (*Salix*) and its secondary hosts are various Apiaceae. On the primary host the fundatrices are a rusty red colour and their offspring are greenish or reddish, with the tips of the antennae and tips of the legs brownish. The siphunculi of wingless females are about twice as long as the cauda, swollen and with a flange. Winged individuals have a black patch on the upper surface of the abdomen, formed by the fusion of the cross bands on segments 3–6. The migration to secondary hosts usually occurs in early summer but this species can be found on willow in August. This species is not attended by ants.

Cavariella archangelicae (Scopoli)

This is a host-alternating species, with willow (*Salix*) as its primary host and angelica (*Angelica* species) as its secondary host. Wingless females are yellowish or green, with a terminal process that is 1.5–2.0 times as long as the base of the last antennal segment. The siphunculi of wingless females are 2.5–2.8 times as long as the cauda, scaly, swollen towards the tip and with a flange. Winged individuals have a dark patch on the upper surface of the abdomen formed by the fusion of the dark green cross bands on segments 3–6. The migration to secondary hosts occurs from May to July. This species is not attended by ants.

Cavariella pastinacae (Linnaeus) **(Pl. 3.4, 3.5)**

This is a host-alternating species, with willow (*Salix*) as its primary host and Apiaceae (mainly hogweed, *Heracleum* and less commonly parsnip, *Pastinaca*) as its secondary hosts. Wingless females are green, with a terminal process that is 3–4 times as long as the base of the last antennal segment. The siphunculi of wingless females are 2.3–3.0 times as long as the cauda, slightly scaly, slightly swollen towards the tip, and with a flange. Winged individuals have a broad dark green to black patch on the upper surface of the abdomen formed by fusion of cross bands on segments 3–6, and the antennae, cauda, supracaudal process and the outer halves of the siphunculi are dark. This aphid can be found on willow throughout summer and is not attended by ants.

Cavariella theobaldi (Gillette & Bragg)

This is a host-alternating species, with willow (*Salix*) as its primary host and Apiaceae (mainly parsnip, *Pastinaca*, hogweed, *Heracleum*, and *Angelica*) as its secondary hosts. Wingless females are green with a terminal process that is 2.1–3.5 times as long as the base of the last antennal segment. The siphunculi of wingless females are 2.0–2.4 times long as the cauda, scaly, cylindrical or tapering from base to tip, not swollen and with a flange. Winged individuals have a dark patch on the upper surface of the abdomen, formed by the fusion of the cross bands on segments 3–6. The antennae are dark, and the outer halves of the siphunculi are brownish. This aphid can be found on willow throughout summer and is not attended by ants.

Genus *Chaitophorus*

There are more than 75 species in this genus. They all feed on Salicaceae, some on poplar (*Populus*), others on willow (*Salix*), often showing a high degree of host specificity. The upper surface is usually more or less completely covered in plates and often bears nodules, little teeth, or fine spinules, sometimes forming a net-like pattern. The cauda is knobbed in most species and the anal plate is not lobed. The siphunculi are short, stump-shaped and with a net-like pattern, at least at the tip.

Chaitophorus leucomelas Koch **(Pl. 3.6)**

This species lives on the undersides of leaves and on young shoots of poplars (mainly black poplar, *Populus nigra*, and related species and hybrids). It is often found in galls induced by other insects. Wingless females vary in colour, being green or yellow, typically with two dark stripes along the sides. The stripes may be divided segmentally. The tips of the antennal segments and siphunculi are dark. The antennae are half the length of the body, and the terminal process is 2.7–3.3 times as long as the base of the last antennal segment. Winged individuals have dark cross bands, which are nearly fused on abdominal segments 3–6, and large marginal plates. The head, thorax, antennae (but not the base of segment 3) and siphunculi are black. Colonies of this aphid are attended by ants in spring.

Chaitophorus populeti (Panzer) **(Pl. 4.1, 4.2, 4.3)**

This species lives on the young shoots and branches of poplars (mainly white poplar, *Populus alba*, and aspen, *P. tremula*). Wingless females are a shiny dark green to black, typically with a paler stripe along the midline of the thorax and the front of the abdomen. The antennae and siphunculi are dark. The antennae are more than half the length of the body, and the terminal process is about twice the length of the base of the last antennal segment. Winged individuals are black with dark bands across the upper surface, and marginal plates. The wing veins are brown-shadowed. Colonies are attended by ants.

Chaitophorus populialbae (Boyer de Fonscolombe) **(Pl. 4.5)**

This species lives on the undersides of leaves of poplars (mainly white poplar, *Populus alba*, and aspen, *P. tremula*). Wingless females are whitish or pale green, often with small green spots. The head and the tips of the antennal segments and tarsi are brownish. The antennae are 0.6–0.9 times the length of the body, and the terminal process is 2.6–3.3 times as long as the base of the last antennal segment. The head, thorax and antennae of winged individuals are black and the abdomen is green or yellow with dark cross bands and lightly pigmented marginal plates. Colonies are attended by ants.

Chaitophorus salijaponicus niger Mordvilko

This species lives on the leaves of various willows (*Salix* species). Wingless females are blackish brown, with a pale ring around the base of the siphunculi. The antennae, legs and cauda are pale. The antennae are a little more than half as long as the body, and the terminal process 2.1–3.0 times as long as the base of the last antennal segment. The cauda is distinctly constricted. Winged individuals are black, with black cross bands, nearly fused, on the abdomen and similarly pigmented marginal plates. Colonies are not usually attended by ants.

Chaitophorus tremulae Koch **(Pl. 4.4)**

This species lives on the undersides of leaves of aspen (*Populus tremula*), often those spun together by lepidopterous larvae, or in galls of other aphids (Pemphiginae). Wingless viviparous females are dark brown to black, often with a pale stripe along the midline. The antennae and siphunculi are dark, and the legs are brown with the hind pair darker. The antennae are half the length of the body, and the terminal process is 2.1–2.8 times as long as the base of the last antennal segment. Winged individuals have a black head, thorax and antennae, a dark green abdomen with black cross bands nearly fused into a solid patch in the middle of the back, and similarly pigmented marginal plates. Colonies are not usually attended by ants.

Chaitophorus truncatus (Hausmann)

This species lives on the leaves of various narrow-leaved willows (*Salix* species). Wingless viviparous females are light green in spring, often with three dark green stripes along the back, and tend to be much darker or blackish in summer. The antennae and siphunculi are black. The antennae are half the length of the body, and the terminal process is 2.2–3.2 times as long as the base of the last antennal segment. Winged individuals have a dark head, thorax, antennae and siphunculi, and a light green abdomen, with separate dark green cross bands. Colonies are not attended by ants.

Chaitophorus vitellinae (Schrank)

This species lives on young shoots, especially those at the base of stems, and on twigs of narrow-leaved willows. Wingless viviparous females are yellowish green with two darker green stripes along the back. The upper surface is rarely darkly pigmented. The antennae are dark at the base and tip. The antennae are more than half the length of the body, and the terminal process is 1.4–2.0 times as long as the base of the last antennal segment. Winged individuals have a black head and thorax and a green abdomen with cross bands and marginal plates. Colonies are attended by ants.

Genus *Chromaphis*

There are two species in this genus and they are both associated with walnut (*Juglans*). All viviparous females are winged. The antennae are shorter than the body. The cauda is knobbed and the anal plate is slightly notched. The siphunculi are short, truncate and without a flange.

Chromaphis juglandicola (Kaltenbach) **Walnut aphid (Pl. 4.6, 5.1)**

This species lives scattered on the undersides of leaves of walnut (*Juglans regia*). Winged viviparous females are yellowish white, without dark markings except in autumn when a pair of brown spots are present on abdominal segments 4 and 5. The antennae are dark brown at the tips and about half the length of the body, and the terminal process is 0.2 times the length of the base of the last antennal segment, or shorter. The siphunculi are conical and smooth, and have two hairs below.

Genus *Clethrobius*

There are three species in this genus and all live on the young branches or twigs of birch (*Betula*) or alder (*Alnus*). All viviparous females are winged. The antennae are shorter than the body. The cauda is knobbed. The siphunculi are short and truncate.

Clethrobius comes (Walker)

This a large species living on branches and twigs, especially those at the base of birch trees, often where the new growth is dying; or twigs of alder overhanging streams. Wingless viviparous females are brown or greenish black. The antennae, legs and cauda are black, but the base of the femur is paler. The antennae are about 0.7 times the length of the body and the terminal process is half the length of the base of the last antennal segment. The cauda is knobbed and distinctly constricted. Colonies are sometimes attended by ants. This aphid is very sensitive to alarm pheromones and drops off the plant when disturbed.

Genus *Colopha*

There are six species in this genus, some of which host alternate between elm (*Ulmus*) and sedges (Cyperaceae).

Colopha compressa (Koch) (Pl. 5.2, 5.3)

This is a host-alternating species, with elm (*Ulmus*) as its primary host and species of sedge (*Carex*) as its secondary host. The fundatrices induce laterally compressed, cockscomb-shaped, yellowish or reddish galls along the mid-rib on the upper surface of elm leaves. On secondary hosts the aphids colonize the roots and they are sometimes found in ants' nests. Fundatrices are yellow or yellowish green and lightly covered with waxy powder. The antennae and legs are brownish. The body is oval or elongate, with wax plates on each of the thoracic and abdominal segments. The antennae are about 0.12 times as long as the body. Small siphuncular pores may be present or absent. Winged viviparous females are green and lack distinct wax glands. The antennae are about 0.25 times the length of the body. There is only one fork in the medial vein of the forewing.

Genus *Corylobium*

There is only one species in this genus and it lives on hazel (*Corylus*). The wingless females are distinguished by having capitate (knob-tipped) hairs on warts.

Corylobium avellanae (Schrank)

A holocyclic species living on the undersides of young leaves and especially the tips of stems of fast growing shoots of hazel (*Corylus avellana*). Wingless viviparous females are a

yellowish green, sometimes mottled with red. The antennae
are pale although the tips of the segments are dark, similarly
the siphunculi, except for their dark tips, and the cauda are
pale. The antennae are 1.3–1.9 times as long as the body, the
terminal process is 4.3–9.7 times as long as the base of the last
antennal segment, and the inner margins of the lateral frontal
tubercles diverge. The siphunculi are 4.2–5.5 times as long as
the cauda, and they taper towards the end and have a net-
like pattern below the well developed flange. The cauda is
short and triangular. Winged individuals have a brown head
and thorax, and a green or dirty red abdomen with cross
bands and marginal plates. The antennae are black and the
siphunculi brownish with a paler base. The species is not
attended by ants.

Genus *Crypturaphis*

There is only one species in this genus and it lives on Italian
alder (*Alnus cordata*). It is possible that in future this species
may also colonize *A. glutinosa*, the alder species native in UK.
The females are distinguished by having peculiar projections
on the head and prothorax.

Crypturaphis grassii Silvestri **(Pl. 5.4, 5.5)**

A holocyclic species living scattered along the veins of leaves
of alder. Wingless viviparous females are dorsoventrally
flattened. The antennae are 5-segmented, and shorter than
the body. Winged individuals have black lobes on the head
and thorax, a paler thorax and a variably developed brown
patch on the upper surface of abdominal segments 5 and 6.

Genus *Drepanosiphum*

There are seven species in this genus and they are all
associated with maples and sycamore (*Acer*). All viviparous
females are winged and live on the undersides of leaves.
Although they are often abundant they never form dense
colonies. The antennae are longer than the body. The cauda is
knobbed, with a more or less distinct constriction, and the
anal plate is almost semicircular. The siphunculi are long,
slightly swollen or almost cylindrical, and smooth, and have
a ring-like constriction below the flange.

Drepanosiphum acerinum (Walker)

A holocyclic species usually living on sycamore saplings
(*Acer pseudoplatanus*) growing in shade. Viviparous females
are yellowish or whitish yellow with a darker thorax. The
abdomen has one or two cross bands in front of the
siphunculi, and usually a small dark spot near the base of

each siphunculus. The segments of the antennae, although pale, have brownish tips, and the only slightly swollen siphunculi have dark, sometimes blackish, tips. The antennae are about twice as long as the body.

Drepanosiphum aceris Koch **(Pl. 5.6, 6.1)**

A holocyclic species usually living on the undersides of the lower leaves of field maple (*Acer campestre*). Viviparous females are yellowish with a darker thorax. The abdomen has two or more dark cross bands in front of the siphunculi and usually a small dark spot near the base of each siphunculus. The segments of the antennae are pale except for the tips of segments 3–5, which are dark, and the siphunculi have dark tips and are sometimes even entirely dark. The antennae are nearly twice as long as the body. The forewing has a dark spot at the tip and another at the outer end of the pterostigma.

Drepanosiphum dixoni Hille Ris Lambers

A holocyclic species usually living on the undersides of the lower leaves of field maple (*Acer campestre*) growing in the shade. Viviparous females are greenish white with a brown head and a darker thorax. The abdomen has two broad, almost black cross bands in front of the siphunculi and a dark spot near the base of each siphunculus. The antennae are pale, except for segments 1, 2 and the base of 3. The siphunculi are pale with brown tips, or sometimes entirely dark, and only slightly swollen. The antennae are twice as long as the body or longer. The forewings lack dark spots at the tip but have a spot at the outer end of each pterostigma. Short-winged (brachypterous) females are common in summer.

Drepanosiphum platanoidis (Schrank) **Sycamore aphid (Pl. 6.2)**

A holocyclic species living on sycamore (*Acer pseudoplatanus*). Viviparous females are light green with a yellowish or brownish head and thorax. The abdomens of the aphids that develop in spring and autumn have five or six dark brown cross bands and marginal plates, including one in front of each siphunculus, which are absent in the generation that matures in summer. The antennae are pale brown or darker, with a paler base to antennal segment 3. The siphunculi are pale with a brown tip. The antennae are twice as long as the body, or longer. The forewing lacks both a dark spot at the tip and one at the outer end of the pterostigma, which is defined by two longitudinal brown stripes.

Genus *Dysaphis*

This is a large genus with more than 100 species in the world. The genus is subdivided into three subgenera, two of which occur in the UK. Most species are holocyclic and host-alternating, with species of Maloideae as the primary hosts and various herbaceous plants as secondary hosts. Some of the non-host-alternating species feed either on Maloideae or on herbaceous plants. Some of the non-host-alternating species have lost the ability to produce oviparous females and males, and have become anholocyclic on herbaceous plants. Most species are attended by ants. On the primary host most adults are rather plump and covered in powdery wax, and in spring induce the leaves to curl or produce galls of different colours. The characteristic features of this genus are a short triangular or 5-sided cauda and the presence of spinal tubercles on the head and abdominal segment 8. The morphological differences between species are small, and overlap in the variation in these differences often occurs.

Dysaphis angelicae (Koch)

This is a holocyclic host-alternating species, with hawthorn (*Crataegus*) as its primary host and angelica (*Angelica sylvestris*) as its secondary host. Cherry-red leaf galls are induced on hawthorn. All females of the second generation are winged and migrate. Fundatrices are greenish grey in colour. The 5-segmented antennae are about 0.3 times the length of the body, with a terminal process that is 1.5–2.0 times as long as the base of the last antennal segment. The siphunculi are 1.5–2.0 times as long as the cauda. Winged viviparous females are greyish red with a well developed black patch on the back but no cross bands on abdominal segments 1 and 2. The antennae are about 0.8–0.9 times the length of the body, with a terminal process that is 3.5–4.0 times as long as the base of the last antennal segment. The siphunculi are 1.7 times as long as the cauda.

Dysaphis apiifolia (Theobald) **Hawthorn–parsley aphid** (Pl. 6.3, 6.4)

This species is subdivided into two subspecies, *apiifolia*, which is anholocyclic on Apiaceae, and *petroselini*, which is a holocyclic host-alternating subspecies, with hawthorn (*Crataegus*) as its primary host and various Apiaceae as its secondary hosts. Cherry-red to crimson curled-leaf galls are induced on hawthorn. All females of the second generation seem to be winged and migrate. Fundatrices are greenish grey in colour. The 5-segmented antennae are about 0.3 times the length of the body, with a terminal process that is 1.8–2.0 times as long as the base of the last antennal segment. The siphunculi are 1.5–1.75 times as long as the cauda. Winged viviparous females are green with a dorsal abdominal patch which is more or less broken along the inter-segmental borders, and no cross bands on abdominal segments 1 and 2. The antennae are about 0.7–0.8 times the length of the body, with a terminal process that is 4.3–5.2 times as long as the base of the last antennal segment. The siphunculi are 1.7–2.0 times as long as the cauda.

Dysaphis crataegi (Kaltenbach) **Hawthorn–carrot aphid**

This species is subdivided into several subspecies: *crataegi*, a holocyclic host-alternating subspecies with hawthorn (*Crataegus*) as its primary host and various Apiaceae, especially carrot, as its secondary host; *aethusae*, a holocyclic host-alternating subspecies with hawthorn as its primary host and various Apiaceae, especially *Aethusa cynapium* and *Torilis japonica*, as its secondary hosts; and *kunzei*, a holocyclic host-alternating subspecies with hawthorn as its primary host and parsnip (*Pastinaca*) as its secondary host. Cherry-red to crimson curled-leaf galls are induced on hawthorn. Nearly all females of the second generation are winged and migrate. Fundatrices are greenish grey and the immature winged females are pinkish. The winged spring migrants of these subspecies differ in the number of hairs on the front part of the sub-genital plate, the shape of the dorsal abdominal patch, and the number of secondary rhinaria on antennal segment 5. Thus for reliable identification large samples need to be analysed.

Dysaphis devecta (Walker) **Rosy leaf-curling aphid (Pl. 6.5, 6.6)**

This is a holocyclic non-host-alternating species, which lives on apple (*Malus*). Fundatrices induce a small gall formed by down-folding of the tip of the lamina of a young leaf. Their offspring leave this gall and induce red roll-galls on the margins of the leaves. The fourth generation, born during summer, consist of sexuales. Fundatrices are bluish grey. The antennae are 5- or 6-segmented, about 0.3 times the length of the body, with a terminal process that is 1.6–2.1 times as long as the base of the last antennal segment. The siphunculi are 1.2–1.6 times as long as the cauda. Winged viviparous females are reddish, with a dark pattern of plates and more or less fused cross bands. The antennae are about 0.6–0.7 times the length of the body, with a terminal process that is 3.5–4.3 times as long as the base of the last antennal segment. The siphunculi are 1.4–1.6 times as long as the cauda.

Dysaphis ranunculi (Kaltenbach)

This is a holocyclic host-alternating species, with hawthorn (*Crataegus*) as its primary host and buttercups (*Ranunculus* species, especially *R. repens* and *R. bulbosus*) as its secondary hosts. Pale yellowish green leaf galls, often suffused with rosy pink, are induced on the primary host. All females of the second generation are winged and migrate. Fundatrices are a dark, blackish green. The 5-segmented antennae are about 0.3 times the length of the body, with a terminal process that is 1.5–2.0 times as long as the base of the last antennal segment. The siphunculi are nearly twice the length of the cauda. Winged viviparous females are a greyish green, with a dorsal four-sided patch, which lacks pale windows or gaps and net-like patterning, and they have a cross band on abdominal segment 2. The antennae are about 0.75 times the length of the body, with a terminal process that is 3.5 times as long as the base of the last antennal segment. The siphunculi are 1.75 times as long as the cauda.

Dysaphis (Pomaphis) aucupariae (Buckton)

This is a holocyclic host-alternating species, with whitebeams (*Sorbus,* mainly wild service tree, *S. tormentalis*) as its primary host and plantain (*Plantago* species) as its secondary host. Yellowish to reddish rolled or twisted pseudo-galls are induced on the leaves of the primary host. The majority of the females of the third generation are winged and migrate. Fundatrices are a mottled greyish and bluish green, with the head, antennae, legs and siphunculi black. The 6-segmented antennae are about 0.3–0.4 times the length of the body, with a terminal process that is 1.3–1.4 times as long as the base of the last antennal segment. Winged viviparous females have black antennae and legs, and an ochreous to greenish yellow abdomen with a black trapeziform dorsal patch on segments 3–5. Abdominal segments 1 and 2 have cross bands, often broken in the middle, there is a cross band on segment 6 fused with the postsiphuncular plates, and there are cross bands on segments 7 and 8. The antennae are about 0.7–0.9 times the length of the body, with a terminal process that is 2.3–3.3 times as long as the base of the last antennal segment. The siphunculi are 1.8–2.4 times as long as the cauda.

Dysaphis (Pomaphis) plantaginea (Passerini) **Rosy apple aphid (Pl. 7.1, 7.2)**

This is a holocyclic host-alternating species, with apple (*Malus*) as its primary host and plantain (*Plantago* species) as its secondary host. Curled and yellow leaves are induced on the primary host. The majority of the females of the third generation are winged and migrate. Colonies may be found on apple as late as July. Fundatrices are a dark reddish to greyish. Their antennae are dark, except for segment 3 and sometimes also 4, and the second and third femora and siphunculi are dark towards the tip. The 6-segmented antennae are about half the length of the body, with a terminal process that is about twice as long as the base of the last antennal segment. Winged viviparous females have very dark antennae and legs, except for the bases of the femora and tibiae, and a reddish grey abdomen, with a black trapeziform dorsal patch on segments 3–5. Abdominal segments 1 and 2 have cross bands, which are sometimes broken in the middle, and segment 6 has a cross band fused with the postsiphuncular plates. The antennae are about 0.95–1.05 times the length of the body, with a terminal process that is 3.9–5.1 times as long as the base of the last antennal segment. The siphunculi are 2–3 times as long as the cauda, which is pointed and triangular.

Dysaphis (Pomaphis) pyri (Boyer de Fonscolombe)
Pear–bedstraw aphid (Pl. 7.3)

This is a holocyclic host-alternating species, with pear (*Pyrus communis*) as its primary host and *Galium* species as its secondary host. Distorted and yellow coloured leaves are induced on the primary host. Winged females of the second or third generation migrate to secondary host plants. Colonies may be found on pear as late as August. Fundatrices are brown. The 5- or 6-segmented antennae are about a third of the length of the body, with a terminal process that is 0.9–1.3 times as long as the base of the last antennal segment. Winged viviparous females are brownish red, with a dorsal patch on segments 3–5. Abdominal segments 2 and 6 to 8 have cross bands. The terminal process is 2.9–4.3 times as long as the base of the last antennal segment. The siphunculi are 1.6–2.6 times as long as the cauda.

Dysaphis (Pomaphis) sorbi (Kaltenbach)

This is a holocyclic, facultatively host-alternating species, with rowan (*Sorbus aucuparia*) as its primary host and bellflowers (*Campanula* species) or sheep's bit (*Jasione*) as its secondary hosts. Curled leaves forming leaf nests are induced on the primary host. Winged females are not produced in large numbers until the fourth generation in June, and colonies may be found on rowan throughout summer. Fundatrices are pale greenish, with dark antennae, brown legs with darker tips to the tibiae and yellowish siphunculi. The 6-segmented antennae are about 0.4 times the length of the body, with a terminal process that is 1.0–1.6 times as long as the base of the last antennal segment. Winged viviparous females have dark antennae and siphunculi and an ochreous to reddish yellow abdomen, with a dark trapeziform dorsal patch, which is broadest on segments 3 and 4. Abdominal segment 1 has spinal plates, segment 2 has a cross band, sometimes broken in the middle, segments 2–7 have marginal plates, which on 6 are fused with the postsiphuncular plates, and segment 7 has a cross band. The antennae are about 0.8–1.0 times the length of the body, with a terminal process that is 2.8–4.0 times as long as the base of the last antennal segment. The siphunculi are cylindrical and slender, and 2.4–3.2 times as long as the cauda, which is triangular.

Genus *Eriosoma*

There are 32 species in this genus, most of which host alternate. On elm the immature fundatrices feed on growing shoots, galling the leaves beyond the feeding site. These leaf galls are then colonized by the fundatrices. Wingless and winged aphids are characterized by having rather conspicuous siphuncular pores with partially chitinized rims surrounded by a ring of hairs. The medial vein in the forewing only branches once.

Eriosoma lanigerum (Hausmann) **Woolly apple aphid (Pl. 7.4, 7.5)**

This is an anholocyclic species, particularly associated with apple (*Malus*), although it sometimes occurs on other plants (*Cotoneaster*, hawthorn and pear). It feeds on the roots, trunks and branches. It overwinters mainly as a first or second instar nymph, in fissures, wounds and hollows in the bark of mainly the lower part of the trunk or branches. Occasionally it overwinters on the roots. Colonies are easily identifiable as they are completely covered by white waxy wool. Wingless viviparous females are purple, red or brown, with distinct wax glands; there are several on the head, and one spinal and one marginal pair on each of the thoracic and abdominal segments. The usually 6-segmented antennae are about 0.17–0.24 times the length of the body. Winged viviparous females are reddish brown and have indistinct very small wax glands, usually one spinal and one marginal pair on each of the abdominal segments. The antennae are about 0.4 times the length of the body, with a terminal process that is a third of the length of the base of the last antennal segment. The siphuncular pores are on dark, low, hairy warts.

Eriosoma patchiae (Börner & Blunck)

This is a holocyclic facultatively host-alternating species, with elm (*Ulmus* species) as its primary hosts and ragwort (*Senecio* species) and *Cineraria* as its secondary hosts. Short shoots and curled, twisted and blistered leaves are induced on the primary host. Winged females are produced in the second and subsequent generations and may migrate in June or July, but colonies may be found on elm throughout summer. Fundatrices are green to dark bluish grey. The usually 6-segmented antennae each have a terminal process that is half the length of the base of the last antennal segment. Siphuncular pores are absent. Wingless viviparous females are yellowish green, or slightly red, with indistinct but large wax glands, extending up the back from each side on the rear abdominal segments. The usually 6-segmented antennae are approximately half the length of the body. Winged viviparous females are green or brownish and have narrow cross bands on the abdominal segments. The antennae are approximately half the length of the body, with a terminal process that is 0.4 times the length of the base of the last antennal segment.

Eriosoma ulmi (Linnaeus) **Elm–currant aphid (Pl. 7.6)**

This is a holocyclic host-alternating species, with elm (*Ulmus* species) as its primary host and currant (*Ribes* species) as its secondary host. Galls, formed by rolling in the edges of a leaf towards the underside, are induced on the primary host. This pseudo-gall turns a yellowish or whitish green. Winged females migrate in June or July. The broadly oval fundatrices, flattened on the underside, are dark green to black and covered with wax. The 6-segmented antennae are 0.18–0.20 times the length of the body, with a terminal process that is a quarter of the length of the base of the last antennal segment. Siphuncular pores are absent. Winged viviparous females are a dark green to bluish grey with irregular, dark cross bands on the abdominal segments. The antennae are about half the length of the body, with a terminal process that is approximately half the length of the base of the last antennal segment. Siphuncular pores are large and on rather dark, low, hairy cones.

Genus *Eucallipterus*

This is a genus with two species. They feed on Tiliaceae and are usually not attended by ants. They are fragile, with long thin legs, and all viviparous females are winged. The antennae are about as long as the body. Viviparous females have truncate siphunculi, a knobbed cauda and a two-lobed anal plate.

Eucallipterus tiliae (Linnaeus) **Lime aphid (Pl. 8.1, 8.2, 8.3)**

A holocyclic species living on the undersides of leaves of lime (*Tilia* species). Viviparous females are pale yellow, or orange, with dark stripes along the sides of the head and first thoracic segment. The abdomen has two rows of dark spots on the back, plus marginal spots on some segments. The antennae are black with the middle part of segment 3, the bases of segments 4, 5 and 6 and the terminal process paler. The siphunculi are usually dark. The 6-segmented antennae have a terminal process that is 0.55–0.7 times the length of the base of the last antennal segment. The forewing has a dark front edge and dark spots at the tips of the oblique veins.

Genus *Euceraphis*

This is a genus with six species, living on birch (*Betula*) and alder (*Alnus*). They are not attended by ants. These aphids

are large and fragile with long thin legs. All viviparous
females are winged and live on young shoots or the
undersides of leaves. Although they often occur in large
numbers they do not aggregate and form dense colonies. The
antennae are usually longer than the body. The cauda is
knobbed and the anal plate rounded.

Euceraphis betulae (Koch) **(Pl. 8.4)**

A holocyclic species living on the undersides of leaves of
Betula pendula. Viviparous females vary from light green to
light yellow, and usually have a black head and thorax. The
abdomen is variously marked with black pigment and
covered with bluish white wax. The upper surface of the
abdomen may be unmarked or have black cross bands (in
spring), or black patches only on segments 3 and 4 (in
autumn).

Euceraphis punctipennis (Zetterstedt) **(Pl. 8.5, 8.6)**

A holocyclic species living on the undersides of leaves of
Betula pubescens. Viviparous females are pale green to light
yellow and have a dark brown head and thorax. The
abdomen is covered with bluish white wax and variably
pigmented. The upper surface of the abdomen may be
unmarked or have black cross bands or black patches only
on segments 4 and 5 or only on 4 (even the spring
generations). The antennae are as long as the body and the
terminal process is shorter than the base of the last antennal
segment. The siphunculi are pale to brownish, truncate and
shorter than the cauda, which is pale or slightly brown.

Genus *Kaltenbachiella*

This is a genus with eight species, which are primarily
associated with elm (*Ulmus*). Like species of the related
genus *Colopha* they also induce galls, but in this case on the
midrib close to the base of a leaf.

Kaltenbachiella pallida (Haliday) **(Pl. 9.1)**

A holocyclic species which host alternates between elm
(*Ulmus*) as the primary host and species of various genera of
Lamiaceae as secondary hosts. The fundatrices induce
globular, closed, pale galls, which are densely covered with
short hairs. On the secondary hosts, the aphids colonize the
roots. The fundatrices are yellow. The body is oval and
elongate, and lacks wax glands. The 4-segmented antennae
are about 0.15 times the length of the body. Siphunculi are
absent. Winged viviparous females are dark green and lack
wax glands. The antennae are 6-segmented, and about 0.4

times the length of the body. The medial vein of the forewing is usually unbranched, or rarely has one fork. Siphunculi are absent.

Genus *Lachnus*

There are 22 species in this genus. They are mostly associated with broadleaved trees, mainly Fagaceae, and are usually attended by ants. They are medium-sized to large, with 6-segmented antennae. More than half the area of the forewings is covered with dark pigment and the medial vein forks twice. The cauda and anal plate are rounded.

Lachnus pallipes (Hartig)

A holocyclic species living on two-year-old or older branches of beech (*Fagus sylvatica*), and on oak (*Quercus*). Wingless viviparous females are a shiny dark brown but for the siphuncular plate, which is relatively pale. The upper surface of the abdomen is densely covered with hairs. The antennae are 0.4–0.5 times the length of the body, and the terminal process is 0.3–0.4 times the length of the last antennal segment. Antennal segment 3 of winged females has 5–7 secondary rhinaria.

Lachnus roboris (Linnaeus) **(Pl. 9.2, 9.3)**

A holocyclic species living on thin, but more than one-year-old branches of oak (*Quercus* species) and sweet chestnut (*Castanea*). Wingless viviparous females are a shiny blackish brown, with large, dark siphuncular cones. The upper surface of the abdomen bears a few short hairs. There are two conical tubercles on the front of the mesosternum. The antennae are 0.4–0.5 times the length of the body, and the terminal process is 0.4–0.45 times the length of the last antennal segment. There are 3–15 secondary rhinaria on the third antennal segment of winged females.

Genus *Monaphis*

There is only one species in this genus. It is large and feeds on birch, and all the viviparous females are winged.

Monaphis antennata (Kaltenbach) **(Pl. 9.4)**

A holocyclic species living mainly on twigs or the upper surface of leaves, often close to where the midrib meets the petiole, on birch (mainly *Betula pendula*). Viviparous females are green, with antennae that are black except at the base. The antennae are longer than the body and the terminal process is nine times as long as the last antennal segment. There are about

40 secondary rhinaria on the third antennal segment of winged females. The siphunculi are very short, with a flange. The cauda is tongue-shaped and not constricted, and the anal plate two-lobed. This aphid is extremely rare and there is usually only one aphid on a leaf. Nymphs usually press themselves close along the midrib and do not react to disturbance.

Genus *Myzocallis*

There are 42 species in this genus. The hosts are deciduous trees and bushes of various families. All viviparous females are winged. The radial sector in the forewing is reduced in some but not all species. The antennae are as long as, or shorter than, the body. The cauda is knobbed and the anal plate two-lobed. The siphunculi are short, truncate and without a flange. These aphids are not attended by ants.

Myzocallis (Agrioaphis) castanicola Baker

A holocyclic species living on the undersides of leaves of sweet chestnut (*Castanea sativa*) and occasionally oak (*Quercus* species). Viviparous females are yellow, with a dark stripe along the middle of the head and thorax, paired black spots on the abdomen and dark siphunculi. The antennae are dark, except for antennal segment 1 and the base of segments 3 and 4. The terminal process is 1.5–2.0 times as long as the base of the last antennal segment. The radial sector in the forewing is well developed.

Myzocallis boerneri Stroyan **Turkey oak aphid (Pl. 9.5, 9.6)**

A holocyclic species living on the undersides of leaves of oak (mainly *Quercus cerris*). Viviparous females are pale yellowish, with small paired brown spinal and marginal spots on the abdomen, and brown-black rings on the antennae. The radial sector in the forewing is well developed.

Myzocallis carpini (Koch)

A holocyclic species living on the undersides of leaves of hornbeam (*Carpinus betulus*). Viviparous females are yellowish white with a pale abdomen and usually unpigmented spinal and marginal plates. The tips of antennal segments 4 and 5 and the base of 6 are dark. The terminal process is 2.3–3.0 times as long as the base of the last antennal segment. The forewing has a reduced radial sector and a black spot at the base of the pterostigma.

Myzocallis coryli (Goeze) **Hazel aphid (Pl. 10.1, 10.2)**

A holocyclic species living on the undersides of leaves of hazel (*Corylus avellana*). Viviparous females are yellowish white with a pale abdomen and usually unpigmented spinal and marginal plates. The tips of antennal segments 4 and 5 and the base of 6 are dark. The terminal process is 2.2–2.7 times as long as the base of the last antennal segment. The forewing has a weakly developed radial sector and a black spot at the base of the pterostigma.

Genus *Myzus*

This is a large genus, with more than 50 species. The morphological differences between species are small, but the species differ greatly in their host plants. The primary hosts of the host-alternating species are mainly species of *Prunus*, but the secondary hosts belong to various families. Most species are small to medium-sized, with converging antennal tubercles. Winged females have dark cross bands, which are fused in most species on segments 3–5. The siphunculi are cylindrical or slightly swollen towards the end, and in most species they have a well developed flange. The cauda is rather short, and triangular, 5-sided or tongue-shaped.

Myzus cerasi (Fabricius) **Cherry aphid (Pl. 10.3, 10.4)**

This is a holocyclic, host-alternating species, with a species of *Prunus* as its primary host and *Galium, Euphrasia, Veronica* and some other herbaceous plants as secondary hosts. This aphid induces the leaves to curl and produce leaf nests on the primary host. Colonies may be found on cherry throughout summer. Fundatrices are shiny and very dark. The 5-segmented antennae are about 0.2 times the length of the body, with a terminal process that is 1.1–1.8 times as long as the last antennal segment. Wingless viviparous females are shiny black or dark brown, with black siphunculi, and a brown cauda, and the bases and tips of the antennae dark. The antennae are about 0.7–0.8 times the length of the body, with a terminal process that is 2.2–4.5 times as long as the base of the last antennal segment. The siphunculi are cylindrical, curving outwards towards the tip, 2.7–3.3 times as long as the triangular cauda and with a large flange.

Myzus padellus Hille Ris Lambers and Rogerson

This is a host-alternating species, with bird cherry (*Prunus padus*) as its primary host and Lamiaceae (*Galeopsis* species) and Scrophulariaceae as secondary hosts. Sexuales of this species have not been recorded. This aphid induces red and yellow leaf galls on the primary host. Wingless viviparous females are yellow, with the head, siphunculi and cauda dark. The antennae are about 0.6–0.7 times the length of the body, with a terminal process that is 2.6–3.0 times as long as the base of the last antennal segment. The siphunculi are cylindrical, scaly, and 2.2–2.7 times as long as the triangular cauda. Winged viviparous females have a yellow abdomen, with a dark patch on segments 3-5, with smaller plates in front of and behind the patch, and marginal and postsiphuncular plates. The antennae are about 0.9 times the length of the body, with a terminal process that is 2.9–3.4 times as long as the base of the last antennal segment. The siphunculi are 1.9–2.5 times as long as the cauda.

Genus *Ovatus*

There are 14 species in this genus. They are morphologically similar to *Myzus*, but the winged females do not have a central black patch on the upper surface of the abdomen. Some species host alternate between hawthorn (*Crataegus*) or other woody Maloideae and herbaceous plants belonging to the Lamiaceae, whereas others do not host alternate and are associated with herbs belonging to various families. Most species are small to medium-sized, with well developed antennal tubercles with round projections on their inner margins. The siphunculi are cylindrical, and often slightly curved, with a distinct flange. The cauda is rather short, and triangular, 5-sided or tongue-shaped. These aphids are not attended by ants.

Ovatus crataegarius (Walker) **Mint aphid**

This is a holocyclic, host-alternating species, with hawthorn (*Crataegus* species) and other Maloideae as its primary hosts and mint (*Mentha* species) and some other Lamiaceae as secondary hosts. On the primary host the aphid lives on the undersides of young leaves and does not deform the leaves. Fundatrices are light green. The antennae are about 0.8–1.0 times the length of the body, with a terminal process that is 3.6–4.6 times as long as the last antennal segment. The siphunculi are 2.1–2.6 times as long as the cauda. Wingless viviparous females are shiny yellowish green to mid-green, with pale siphunculi and cauda. The antennae are curved, about 1.2–1.5 times as long as the body, with a terminal process that is 5.2–7.6 times as long as the base of the last antennal segment. The siphunculi are 1.7–2.6 times as long as the tongue-shaped cauda.

Ovatus insitus (Walker)

This is a holocyclic, host-alternating species, with hawthorn (*Crataegus* species) or medlar (*Mespilus*) and occasionally also other genera within the Maloideae as its primary hosts and gypsywort (*Lycopus*) as its secondary host. The aphids colonize the tips of the shoots of the primary host. Fundatrices

are similar to those of *O. crataegarius*. The Siphunculi are
2.0–2.5 times as long as the cauda. Wingless viviparous
females are similar to those of *O. crataegarius*, a shiny greenish
white. The antennae are about 0.9–1.5 times the length of the
body, with a terminal process that is 5.2–7.6 times as long as
the base of the last antennal segment. The siphunculi are
2.1–2.9 times as long as the tongue-shaped cauda.

These two species have different secondary host
plants, but meet on the same primary host. Hybridization
may be prevented by species specific attraction of the males
to their sexual females (p. 7).

Genus *Pachypappa*

There are more than ten species in this genus. Most of them
host alternate. They overwinter on poplar. This genus is
characterized by fundatrices that lack wax glands, and
winged spring migrants that have forewings in which the
medial vein has only one branch.

Pachypappa tremulae (Linnaeus)

A holocyclic species, which host alternates between poplars
(mainly aspen, *Populus tremula*), its primary host, and spruce
(*Picea abies*), its secondary host. The fundatrices feed on the
stems of twigs and their offspring colonize the new shoots as
they emerge from the buds and induce the petioles and leaves
to curl, forming a leaf nest. These leaf nests are visited by
ants. All second generation aphids are winged and migrate.
Fundatrices are reddish brown, but appear silvery as they are
covered with long fine hairs. The body is globular, with a flat
lower surface. The 5-segmented antennae are about 0.10–0.12
times the length of the body, with a terminal process that is
0.10–0.11 times the length of the base of the last antennal
segment. Siphuncular pores are absent. Winged viviparous
females are orange or reddish brown, and covered with wax.
There are no wax glands on the head. The antennae are 6-
segmented, and about one third of the length of the body,
with a terminal process that is 0.11–0.15 times the length of
the base of the last antennal segment. The siphuncular pores
are very small and often not visible.

Genus *Panaphis*

The two species in this genus are both associated with
walnut (*Juglans*). All viviparous females are winged. The
antennae are shorter than the body. The cauda is knobbed
and the anal plate is slightly notched. The siphunculi are
short, truncate and without a flange.

Panaphis juglandis (Goeze) **Large walnut aphid (Pl. 10.5)**

Small colonies of this holocyclic species typically feed along
the midrib on the upper surface of leaves of walnut (*Juglans
regia*). Winged viviparous females are yellowish white, with
dark bands on the abdomen. In autumn there are cross bands
and paired brown spots on abdominal segments 4 and 5. The
tips of the antennae are dark brown. The antennae are about
half the length of the body and the terminal process is 0.2
times the length of the base of the last antennal segment, or
shorter. The siphunculi are conical and smooth, with two
hairs on the lower surface.

Genus *Patchiella*

There is only one, large, species in this genus.

Patchiella reaumuri (Kaltenbach) **(Pl. 10.6, 11.1, 11.2)**

A holocyclic species, which host alternates between lime
(*Tilia*), its primary host, and Araceae, its secondary hosts. The
fundatrices and their offspring can be found in large leaf-nest
galls induced by the twisting and stunting of the terminal
growth of young shoots on the primary host. These leaf nests
are visited by ants. All the second generation aphids are
winged and migrate. Fundatrices are globular and greenish
to yellowish brown.

Genus *Pemphigus*

This is a large genus comprising more than 70 species. Many
of the species host alternate, inducing galls on leaves,
petioles, or branches of poplar (*Populus*), the primary host.
The morphological differences between species are small, but
the species differ greatly in their biological characteristics.
Most are small to medium-sized. Fundatrices have spinal,
pleural and marginal wax glands on most body segments.
The winged viviparous females that emerge from the galls
have a black head and thorax and a rather elongate yellowish
or greyish green abdomen, which is dusted with wax, and
they do not have wax glands on the head and mesonotum.
On the primary host these aphids are not attended by ants.

Pemphigus bursarius (Linnaeus) **Poplar–lettuce aphid
(Pl. 11.3, 11.4)**

This is a holocyclic, host-alternating species, with poplars
(mainly *Populus nigra*) as the primary hosts and various
herbaceous plants, especially Asteraceae, as secondary hosts.
It is a pest of lettuce. Purse-shaped galls, yellowish or
reddish when mature, are induced on the petioles of the
leaves of the primary host. There can be more than one gall

per petiole, and the leaf lamina may curl and yellow. In summer winged females leave the gall via an opening on the side. Fundatrices are green, with brown head and legs. The 4-segmented antennae are about 0.12–0.15 times the length of the body. These aphids do not have siphunculi. The winged viviparous females that develop on the primary host are greyish green or greyish brown and lightly covered with waxy powder. The usually 6- (rarely 5-) segmented antennae are about 0.33–0.40 times the length of the body and have a distinct terminal process, which is 0.17–0.20 times the length of the base of the last antennal segment. Siphuncular pores are present but small. There is a brown shadowing around each wing vein.

Pemphigus gairi Stroyan

This is a holocyclic, host-alternating species, with poplar (*Populus nigra*) as its primary host and fool's parsley (*Aethusa cynapium*) as its secondary host. Pouch-shaped galls are induced on or near the midrib on the upper surface of the leaves of the primary host. In summer winged females leave the gall via an opening on the underside of the leaf. Fundatrices are pale green, with brown head and legs. The 4-segmented antennae are about 0.12–0.15 times the length of the body. This morph does not have siphunculi. The winged viviparous females that develop on the primary host are yellowish green.

Pemphigus passeki Börner **(Pl. 11.5, 11.6)**

This is a holocyclic host-alternating species, with poplar (mainly *Populus nigra*) as its primary host and caraway (*Carum carvi*) as its secondary host. This aphid induces the basal half of the midrib of the leaves of its primary host to swell and form a gall, which is wedge-shaped with the broadest end close to the petiole. In summer the winged migrants leave the galls through a gap in the gall on the under surface of the leaf. Fundatrices are reddish or greyish green.

Pemphigus populinigrae (Schrank) **(Pl. 12.1)**

This is a holocyclic host-alternating species, with poplars (*Populus* species) as the primary host and cudweeds, *Gnaphalium* and *Filago*, as secondary hosts. Pouch-shaped galls, yellowish or dull reddish when mature, are induced on the midrib of the upper surface of leaves of the primary host. The central part of the gall is smooth and round or oval. In summer winged migrants leave the gall through a gap in the gall on the under surface of the leaf. Fundatrices are green or greyish green, with 4-segmented antennae, which are about 0.17 times the length of the body, and they lack siphunculi. The winged migrants from the primary host are dark green with a slight covering of powdery wax. Along the upper surface of the abdomen there are six rows of more or less fused wax glands. The usually 6- (rarely 5-) segmented antennae are about 0.3–0.4 times the length of the body, with a rather indistinct terminal process. Siphunculi are present but small. There is no brown shadowing around the veins on the wings.

Pemphigus protospirae Lichtenstein **(Pl. 12.2)**

This is a holocyclic, host-alternating species, with poplar (mainly *Populus nigra*) as its primary host and aquatic Apiaceae as secondary hosts. This aphid induces the petiole of the leaves of its primary host to swell, flatten and spiral to form smooth galls, which are green, or green mottled with red, when mature. The sizes and shapes of the galls vary. All the second generation aphids are winged and leave the galls during late spring to early summer. Fundatrices have 4-segmented antennae, and spinal, pleural and marginal wax glands on most body segments. The antennae are about 0.2 times the length of the body. The aphids lack siphunculi. Winged migrants from the primary hosts are greyish green, and lightly covered with waxy powder. There are no wax glands on the middle part of abdomen. The usually 6- (rarely 5-) segmented antennae are about 0.33–0.40 times the length of the body, with a distinct terminal process that is 0.17–0.20 times the length of the base of the last antennal segment. Siphuncular pores are present but small. This species is morphologically similar to *P. bursarius*.

Pemphigus spyrothecae Passerini **(Pl. 12.3)**

This is a holocyclic, non-host-alternating species living on poplar (mainly *Populus nigra*). It induces the petioles of the leaves of its host to swell, flatten and spiral to form smooth galls, which are green, or green mottled with red, when mature. The galls are similar to those of *P. protospirae*, but usually thicker and with fewer spirals. Winged sexuparae leave the gall in autumn. Fundatrices are green, with brown head and legs. The 4-segmented antennae are about 0.12–0.15 times the length of the body. The aphids lack siphunculi. Winged sexuparae are green and covered with powdery wax. Wax glands are present on the thorax and abdomen, which has both marginal and spinal plates. The 6-segmented antennae are about a third of the length of the body, with a terminal process that is about a quarter of the length of the base of the last antennal segment.

Genus *Periphyllus*

There about 40 species in this genus. Most of them are associated with maples (*Acer*) and some with horse chestnut (*Aesculus*). These aphids are medium-sized, elongate oval or pear-shaped. The upper surface of wingless aphids is mainly membranous with small, hair-bearing plates. The antennal terminal process is usually long, and the cauda is broadly rounded, slightly knobbed or tongue-shaped, with a more or less distinct constriction. The siphunculi are stump-shaped, usually with polygonal net-like patterning below the pronounced flange. Colonies of these aphids are usually attended by ants. The association with their host plants is very strong. In summer, when the leaves are mature, some species survive only as small aestivating nymphs. In some species these nymphs are flat and armoured with hardened plates on the back. Leaf-shaped hairs are present around the head and body, and on parts of the legs and antennae. In other species these nymphs are covered with very long hairs. A few species do not produce aestivating nymphs, while others produce both aestivating and normal nymphs during summer.

Periphyllus acericola (Walker) **(Pl. 12.4)**

This is a holocyclic, non-host-alternating species, which lives on sycamore (*Acer pseudoplatanus*), on the under surface of young leaves, petioles and young shoots. Wingless viviparous females are green or yellowish, occasionally with brown markings. The head, pronotum, legs and siphunculi are pale and the tips of the antennal segments are dark. The antennae have a terminal process that is 2.3–3.0 times as long as the last antennal segment. The width of the base of the cauda is more than twice the length of the cauda, and the siphunculi are 2.1–2.6 times as long as the cauda. Winged viviparous females have broad, dark dorsal cross bands, pale and indistinct marginal plates, dark siphunculi and a very dark pterostigma. Their antennae have a terminal process that is 2.4–2.9 times as long as the base of the last antennal segment. The aestivating nymphs are yellowish white, with long pointed hairs about half as long as the body. The clusters of aphids look like whitish spots on the under surfaces of the leaves.

Periphyllus aceris (Linnaeus) **(Pl. 12.5)**

This is a holocyclic, non-host-alternating species, which lives on maple (*Acer platanoides*), on the under surface of young leaves, petioles and young shoots. Wingless viviparous females are yellow, occasionally with green markings. The head, pronotum, legs (except tarsi) and siphunculi are pale and the tips of the antennal segments are dark. The antennae have a terminal process that is 2.2–2.7 times as long as the last antennal segment. The cauda is distinctly shorter than the width of the base, and the siphunculi are 2.1–2.5 times as long as the cauda. Winged viviparous females have broad, narrow, black dorsal cross bands, and equally dark marginal plates, dark siphunculi and a very dark pterostigma. Their antennae have a terminal process that is 1.9–2.5 times as long as the base of the last antennal segment. The aestivating nymphs look like those of *P. acericola*, but they differ in the shape of some hairs on the second tarsal segments.

Periphyllus hirticornis (Walker)

This is a holocyclic non-host-alternating species that lives on field maple (*Acer campestre*), where it colonizes the under surfaces of young leaves, leaf petioles and developing seeds. Wingless viviparous females are light green, without dark markings. The antennae have a terminal process that is five times as long as the last antennal segment. The cauda is slightly knobbed and has a narrow base. Winged viviparous females have variably developed plates, no dorsal cross bands and brownish siphunculi and cauda. Antennal segments 1, 2 and 6, and the tips of 3–5, are dark. The antennae have a terminal process that is about five times as long as the base of the last antennal segment. The cauda is similar in shape to that of the wingless viviparous female. Colonies are often attended by ants. The aestivating nymphs are light green, with leaf-like marginal hairs and red eyes.

Periphyllus obscurus Mamontova

This non-host-alternating species lives on the under surfaces of leaves, and on petioles and young shoots, of field maple (*Acer campestre*). Wingless viviparous females are blackish green, with black siphunculi. The antennae have a terminal process that is 3.2–4.1 times as long as the last antennal segment. The cauda is broadly rounded. It is unknown whether this species produces aestivating nymphs.

Periphyllus testudinaceus (Fernie) **(Pl. 12.6)**

This is a holocyclic non-host-alternating species that lives on the leaves, petioles and young shoots of several species of maple (especially *Acer campestre* and *A. pseudoplatanus*) and sometimes also horse chestnut (*Aesculus* species). Wingless viviparous females are green or dark brown. Antennal segments 1, 2 and 6, and the tips of 3–5, are dark. The antennae have a terminal process that is 2.5–3.7 (2.1 in autumn) times as long as the last antennal segment, and the cauda is twice as broad as long. Winged viviparous females have dark marginal plates, broad dorsal cross bands that are darker than the pterostigma, and brownish siphunculi and cauda. Antennal segments 1, 2 and 6, and the tips of 3–5, are dark. The antennae have a terminal process that is about five times as long as the base of the last antennal segment. The shape of the cauda is as in the wingless viviparous female. Colonies are often attended by ants. The aestivating nymphs are light green, with leaf-like marginal hairs and red eyes.

Genus *Phyllaphis*

The two species in this genus live on beech (*Fagus*). The characteristic features of this genus are abdominal segments each with one spinal, one pleural and one marginal pair of honeycomb-like wax glands, and a terminal process that is much shorter than the base of the last antennal segment. The cauda is knobbed and the siphunculi consist of just pores.

Phyllaphis fagi (Linnaeus) **(Pl. 13.1, 13.2)**

This is a holocyclic non-host-alternating species that lives on the undersides of young leaves of beech (*Fagus*), which often bend downwards along the midrib. Wingless viviparous females are elongate oval, pale yellowish green and covered with waxy wool. The wax glands are pale or dark. The antennae are a little shorter than the body, with a terminal process that is 0.11–0.12 times the length of the last antennal segment. The anal plate is rounded and very slightly notched. Winged viviparous females are greenish, with wax glands on dark marginal and variably developed dorsal plates and a dark head and thorax with wax glands. The antennae are about 0.75 times the length of the body with a terminal process that is 0.10–0.13 times the length of the base of the last antennal segment. The cauda is dark and has a globular end, and the anal plate is notched. Colonies are not attended by ants.

Genus *Plocamaphis*

This is a small genus with only five species, which all live on willow (*Salix*). They are related to *Pterocomma*, but differ in being less hairy. The siphunculi are club-shaped, with a thin basal part and a rather small un-flanged opening in the centre of the rounded tip. These aphids secrete flocculent wax and are not attended by ants.

Plocamaphis amerinae (Hartig)

This is a holocyclic, non-host-alternating species, which lives on the shoots of many species of willow (especially *Salix alba* and *S. purpurea*). Wingless viviparous females are yellowish, greenish or brownish. The head, pronotum, antennae and legs are dark, and the siphunculi are a very pale yellowish or orange. There are large dorsal plates on abdominal segments 1–2 and very small ones on segments 3–5. The antennae have a terminal process that is 1.5–1.75 times as long as the last antennal segment. Winged viviparous females have a dark brown head and thorax, and a paler brown abdomen with inter-segmental plates.

Genus *Prociphilus*

There are 46 species in this genus. Unlike the other genera belonging to the subfamily Pemphiginae, their primary hosts are not poplar but members of the Rosaceae, Caprifoliaceae and Oleaceae. The secondary hosts are mostly coniferous trees. Species in this genus have wax glands on the head, thorax and abdomen, even the fundatrices. The medial vein on the forewing is unbranched and the aphids lack siphunculi.

Prociphilus (Stagona) pini (Burmeister)

This is a holocyclic, host-alternating species, which has hawthorn (*Crataegus* species) as its primary host and pine (*Pinus*) and its close relatives as secondary hosts. This aphid lives on the under surfaces of the young leaves of its primary host, which curl and turn yellow. The second generation are all winged and leave the primary host. Fundatrices are green or brownish green and covered in waxy powder. The 5-segmented antennae are about 0.25–0.3 times the length of the body. Winged viviparous females have a light green or greyish green abdomen. The 6-segmented antennae are about half the length of the body, with a terminal process that is a quarter of the length of the base of the last antennal segment. This aphid lacks siphunculi.

Prociphilus fraxini (Fabricius) **(Pl. 13.3, 13.4)**

The primary host of this holocyclic, host-alternating species is ash (*Fraxinus excelsior*) and its secondary host is fir (*Abies*). The fundatrices colonize the base of the new shoots of the primary host. Their offspring feed on the young shoots and petioles inducing the formation of dense leaf nests, mostly high in the trees. The females of the second generation are all winged and leave the primary host. Fundatrices are brown and covered with white waxy wool. The wax glands on the head and thorax are very distinct whereas those on the abdomen, although large, are indistinct. The 5-segmented antennae are about a third of the length of body, with a terminal process that is 0.2 times the length of the base of the last antennal segment. Winged viviparous females have a blackish brown head and thorax, and a light brown or yellowish red abdomen. The 6-segmented antennae are about half the length of the body, with a terminal process that is 0.15 times the length of the base of the last antennal segment. This species lacks siphunculi.

Genus *Pterocallis*

There are 12 species in this genus and all are associated with Betulaceae, especially alder (*Alnus*). They are small and pale, and usually live scattered on the under surfaces of leaves. The antennae are as long as the body or shorter, with a short terminal process. The siphunculi are short and truncated. In viviparae the cauda is knobbed and the anal plate two-lobed. They are mostly not attended by ants.

Pterocallis alni (de Geer)

This holocyclic, non-host-alternating species lives scattered on the under surfaces of the leaves of alder (*Alnus glutinosa*). The aphids are rarely attended by ants. Wingless viviparous females are yellowish white or greenish white, with the tips of the antennal segments black and the tips of the siphunculi dark. There is a black spot on the outer surface of the hind femur near the knee. The hairs on the upper side are pale. The 6-segmented antennae are about 0.6–0.8 times the length of the body. The terminal process is 0.5–0.8 times the length of the last antennal segment. Winged viviparous females are yellowish white or a shiny greenish white, with the tips of their antennal segments and siphunculi pale. There are 2–5 rather large secondary rhinaria on the basal half of the third antennal segment.

Pterocallis maculata (von Heyden)

This holocyclic, non-host-alternating species lives in colonies, especially along veins, on the under surfaces of leaves of alder (*Alnus glutinosa*), and is attended by ants. Wingless viviparous females are yellowish, with a variably developed pattern of dark green dorsal cross bands, black tips to the antennal segments and dark siphunculi, except for their bases. There is a black spot on the outer surface of the hind femur near the knee. The hairs on the upper side are pigmented. The 6-segmented antennae are about 0.4–0.6 times the length of the body. The terminal process is half the length of the last antennal segment. Winged viviparous females are greenish with dark green markings. There are 2–5 rather large secondary rhinaria on the basal half of the third antennal segment.

Genus *Pterocomma*

The 30 species in this genus are associated with willows, Salicaceae. They are often dark, with contrasting brightly coloured siphunculi. They live in colonies on the branches and twigs and are attended by ants. The convex body has a flat underside and is powdered with wax, especially along the borders of the segments. The body, antennae and legs are densely covered with long pointed hairs. The antennae are

6-segmented, and about half as long as the body. The siphunculi are longer than wide, cylindrical or swollen, and with a flange.

Pterocomma pilosum Buckton **(Pl. 13.5)**

This holocyclic, non-host-alternating species lives on various species of willow (*Salix*). Wingless viviparous females are greyish or brownish, with yellowish siphunculi. There are marginal plates on the abdominal segments and the pleurospinal plates are either discontinuous or absent. The terminal process is 1.0–1.9 times the length of the last antennal segment. The siphunculi are cylindrical. Winged viviparous females have marginal plates on the abdominal segments and more or less fused pleurospinal plates (plates on spinal and pleural area are fused) and cross bands on abdominal segments 2–8. This species has been subdivided into subspecies (*P. p. pilosum*, *konoi* and *sarmaticum*), but further research is needed to validate their status.

Pterocomma populeum (Kaltenbach)

This holocyclic, non-host-alternating species lives on the branches, trunks and two-year-old twigs of various species of poplar (*Populus*). Wingless viviparous females are grey or brownish, with pale siphunculi. There are rather large marginal plates and dorsal plates on the thorax and abdominal segments. The dorsal plates are more or less broken into smaller plates. The terminal process is 1.3–1.7 times as long as the last antennal segment. The siphunculi are subcylindrical. Winged viviparous females have marginal plates and broad cross bands on the abdominal segments. Some of these bands are incomplete in the middle.

Pterocomma rufipes (Hartig)

This holocyclic, non-host-alternating species lives on the twigs and young branches of various species of willow (*Salix*). It has been found on poplar (*Populus*) in Scandinavia. Wingless viviparous females are grey or dull reddish brown to dark brown, with waxy powder spots and yellowish siphunculi. There are large marginal plates on all the abdominal segments, pairs of large pleurospinal plates on segments 6 and 7 and a cross band on segment 8. The terminal process is 1.3–2.2 times as long as the last antennal segment. The siphunculi are swollen. Winged viviparous females have marginal plates and paired, rather large dorsal plates or cross bands on all abdominal segments.

Pterocomma salicis (Linnaeus) **(Pl. 13.6)**

This holocyclic, non-host-alternating species lives in dense colonies, especially on two-year-old twigs of various species of willow (*Salix*). It has been found on poplar (*Populus*) in Scandinavia. Wingless viviparous females are greenish black to black, with waxy red powder spots and bright red or orange siphunculi. There are marginal plates on all abdominal segments and pleurospinal plates on segments 7 and 8. The terminal process can be either as long as the last antennal segment or shorter. The siphunculi are strongly swollen. Winged viviparous females have abdominal plates as in wingless females.

Pterocomma tremulae Börner

This non-host-alternating species lives on suckers and two-year-old twigs of poplars (mainly aspen, *Populus tremula*). Oviparous females and males have not been found. Wingless viviparous females have pale yellowish siphunculi and are an olive brown, dusted with powdered wax. There are large marginal and pleurospinal plates on the abdominal segments. The terminal process is 1.5–2.0 times as long as the last antennal segment. The siphunculi are slightly swollen, often with 1–3 hairs. The abdominal segments of winged viviparous females have marginal and paired rather large dorsal plates, which usually do not form complete cross bands.

Genus *Rhopalosiphum*

There are about 13 species in this genus. All of them host alternate between *Prunus* or other Rosaceae as primary hosts and sedges (Cyperaceae), grasses (Poaceae), and rarely other plants, as secondary hosts. There are weakly developed tubercles on the frons. The antennae are 5- or 6-segmented and shorter than the body, and the terminal process is longer than the base of the last antennal segment. The siphunculi are longer than the cauda, often slightly swollen, and distinctly constricted below the well developed flange. Cuticular sculpturing is usually well developed on the back, forming a polygonal net-like pattern. Viviparous females have a finger- or tongue-shaped cauda.

Rhopalosiphum insertum (Walker) **Apple–grass aphid (Pl. 14.1, 14.2)**

A holocyclic, host-alternating species, with Maloideae (especially apple (*Malus*), hawthorn (*Crataegus*) and *Cotoneaster*) as primary hosts and various grasses as secondary hosts. The first generation induce the leaves of the primary hosts to curl. Winged females migrate to secondary hosts at the end of May, but colonies can be found on the primary host as late as July. The fundatrices are a light green and have pale siphunculi with dark tips. The fundatrices are fatter than the wingless viviparous females of the following generations. The siphunculi are one tenth as long as the body. The 5-segmented antennae are about a third of the length of the body, with a terminal process that is 2.9 times as long as the base of the last antennal segment. The colonies are attended by ants. Winged viviparous females have a blackish head, thorax and siphunculi. The abdomen is green with brown marginal and postsiphuncular plates and brownish pigmentation on abdominal segments 6–8. The antennae are 6- or (in autumn) 5-segmented. The siphunculi are 1.5–1.8 times as long as the cauda.

Rhopalosiphum nymphaeae (Linnaeus) **Water lily aphid** **(Pl. 14.3)**

A holocyclic, host-alternating species, with cherry (*Prunus* species) as its primary host and various water plants as secondary hosts. The aphids live on young twigs, petioles and fruit stalks of the primary host. Winged females migrate to secondary hosts in May or June. The wingless viviparous females are reddish brown to dark olive, and slightly powdered with grey wax, especially on the upper surface of abdominal segments 1–4. The siphunculi are green or greenish black, and about 0.2 times the length of the body. The 6-segmented antennae are about 0.6 times the length of the body, with a terminal process that is 2.3–3.9 times as long as the base of the last antennal segment. The colonies are attended by ants.

Rhopalosiphum padi (Linnaeus) **Bird cherry–oat aphid** **(Pl. 14.4)**

A holocyclic, host-alternating species, with bird cherry (*Prunus padus*) as its primary host and various grasses as secondary hosts. Viviparae can hibernate on secondary host plants in regions with mild winters. On bird cherry they live on the undersides of young leaves and induce rolls or folds along the length of the leaf, which loosely enclose the colonies. Winged females migrate to secondary hosts in June, but some colonies can be found on bird cherry as late as July. The fundatrices are light green with a reddish spot at the base of each siphunculus. The siphunculi are almost cylindrical and pale, with dark tips. The 5- or 6-segmented antennae are less than half the length of the body, with a terminal process that is less than twice the length of the base of the last antennal segment. The wingless viviparous females are dark green, or brown to almost black, and powdered with wax. The siphunculi are dark, about 0.12–0.14 times the length of the body, about twice as long as the cauda and slightly swollen near the tip. The 6-segmented antennae are about half the length of the body, with a terminal process that is 3–4 times as long as the base of the last antennal segment.

Genus *Stomaphis*

There are 25 species in this genus. All are very large and feed on the stems or roots of their host plants. Whereas the females characteristically have a very long rostrum, the minute males lack mouthparts (they are arostrate) and have a reduced number of larval stages. The long rostrum allows the females to penetrate the thick bark and reach the phloem on the trunks of broadleaved trees. The siphunculi consist of pores on low, hairy cones. The antennae are 6-segmented and densely covered with hairs. These aphids are always attended by ants, which keep the eggs of the aphids in their nests during winter.

Stomaphis longirostris (Fabricius) **(Pl. 14.5, 14.6)**

This holocyclic, non-host-alternating species lives on willows (*Salix* species) and, in continental Europe, on poplars (*Populus* species). The aphids can be found in deep crevices in the bark of the trunks of their host plants. The attendant ants usually encase the aphid colonies in a tunnel of soil and plant material and aggressively defend the aphids against attack by predators and parasitoids. Wingless viviparous females are elongate oval, whitish yellow, and covered with greyish white waxy powder. The rostrum is nearly twice as long as the body. The cauda is rounded. Relative to their body length, winged females have relatively small wings.

Stomaphis quercus (Linnaeus) **(Pl. 15.1, 15.2)**

This holocyclic, non-host-alternating species lives mainly on oak (*Quercus* species). The aphids can be found feeding on the branches and deep in crevices in the bark of the trunks of large oak trees. The aphid colonies are attended and vigorously defended by ants (*Lasius fuliginosus*). Wingless viviparous females are elongate oval, dark green to shiny brown, with dark spots. The antennae are dark, with segment 3 paler than the others. There are dark spinal spots on abdominal segments 1–4, and cross bands, incomplete in the middle, on segments 7 and 8. The antennae are about 0.4 times the length of the body. The cauda is rounded. The rostrum is nearly twice as long as the body. Relative to their size they have small wings, the veins of which have brown borders.

Genus *Symydobius*

The eight species in this genus live on either birch (*Betula*) or alder (*Alnus*). They are medium-sized to large and shiny brown. Wingless and winged females have similar patterns of hardening and pigmentation. They live on branches and twigs and are attended by ants.

Symydobius oblongus (von Heyden)

A holocyclic, non-host-alternating species living in colonies on 3–4-year-old branches and twigs of birch (mainly *Betula pendula* and *B. pubescens*). The aphids are attended by ants, and when disturbed they do not drop off but run away. Their strongly developed tarsal claws allow them to run on rough surfaces. The wingless viviparous females are a shiny brown and do not have a covering of powdery wax. The antennae are brown, but the basal parts of segments 4 and 5, and sometimes also 6, are white. The siphunculi are paler than the body and the semicircular cauda. The siphunculi are short and truncate, and have rows of fine spinules around them. The antennae are a little shorter than the length of the body, with a terminal process that is shorter than the base of the last antennal segment. Winged females have either well developed or short wings.

Genus *Tetraneura*

The 33 species in this genus are primarily associated with elm (*Ulmus*). All legs of wingless females have only one tarsal segment. The winged females have forewings with unbranched medial veins. Wax plates can be either present or absent.

Tetraneura ulmi (Linnaeus) **Elm–grass root aphid (Pl. 15.3, 15.4)**

A holocyclic, host-alternating species with elm (*Ulmus*) as its primary host and various grasses (Poaceae) as secondary hosts. Anholocyclic overwintering occurs on roots of grasses. The fundatrices induce bean-shaped, stalked, smooth, shiny, green or yellowish galls. All females of the second generation are winged and leave the gall through a secondary opening on the side. They migrate to grass, where they colonize the roots. Fundatrices are light green, with the head and thorax dark and the antennae and legs brownish. The body lacks wax glands. The 3- or 4-segmented antennae are about 0.15 times the length of the body. Winged viviparous females have a shiny black head, thorax, antennae and legs, and greyish black abdominal segments. Wax glands are present on the head, thorax and abdomen. The antennae are 6-segmented, about 0.33 times the length of the body, with a terminal process that is about 0.25 times the length of the base of the last antennal segment. Siphuncular pores can be present or absent.

Genus *Thecabius*

There are about 17 species in this genus. Most of them host alternate, inducing galls on leaves, petioles, or branches of poplar (*Populus*), the primary host. The morphological differences between species are very small and in appearance they are very similar to *Pemphigus*. Although very similar morphologically the species differ more markedly in their biological characteristics. All morphs have wax glands. The medial vein in the forewing is unbranched.

Thecabius affinis (Kaltenbach) **(Pl. 15.5, 15.6)**

This is a holocyclic, host-alternating species, which has poplar (*Populus nigra*) as its primary host and creeping jenny (*Lysimachia nummularia*) as its secondary host. Galls are induced by fundatrices, which cause part of the edge of a leaf to bend sharply downwards. Their offspring leave this gall, move to the midrib of a young leaf, and induce the lamina of the leaf to fold along the midrib towards the underside. The upper surface of this roof-like gall develops blisters and gradually turns reddish. The winged females that occur in the second and following generation migrate to the secondary host plant. The body segments of all the different morphs have wax glands, with the exception of winged viviparous females, which do not have wax glands on the head, but do have them on the mesonotum. The fundatrices are green or bluish green and covered with wax, and they lack siphunculi. The winged viviparous females are greenish and covered with wax, and have siphuncular pores. The antennae are about half the length of the body, with a terminal process that is 0.25 times of the length of the base of the last antennal segment.

Thecabius (Parathecabius) lysimachiae Börner

This is a holocyclic, host-alternating species, which has poplar (*Populus nigra*, rarely 'Italica') as its primary host and buttercups (*Ranunculus* species) as its secondary hosts. Oval, thick-walled, pocket-shaped galls are induced by fundatrices generally all over the upper surfaces of the leaves of their primary host. The second generation individuals migrate and induce their own galls by causing the leaves of poplar to fold and become convoluted in an irregular way. The body segments of all the different morphs have wax glands except the head of viviparous females and mesonotum of winged viviparous females.

Genus *Thelaxes*

The four species in this genus all live on oak. They characteristically have a long and almost needle-like last rostral segment. The cauda is knobbed, the siphunculi are virtually pore-like and the anal plate is entire. Wingless females lack compound eyes. The winged females produced early in the year give birth to the males and sexual females, which spend the summer aestivating as very small nymphs and only resume and complete their development in autumn.

Thelaxes dryophila (Schrank) **(Pl. 16.1)**

This is a holocyclic, non-host-alternating species, which lives on various species of oak. The fundatrices are dark, and their antennae are about 0.25–0.33 times the length of the body, with a terminal process that is 0.15–0.20 times the length of the last antennal segment. After the fundatrices become adult ant-attended colonies are found on shoots, young stems, petioles, and the undersides of leaves, especially along the midribs. Wingless viviparous females are broadly oval and dark brownish with a yellowish green stripe along the back. The antennae, legs, siphunculi and cauda are brownish. The 5-segmented antennae are a little less than half as long as the body, with a terminal process that is less than half the length of the base of the last antennal segment. There are two kinds of hairs on the body: fine, and spine-like. Winged females have a black head and thorax, and the antennae, legs and cauda and the areas around the siphuncular pores are dark. When at rest this aphid, unlike all other winged aphids, folds its wings horizontally over its abdomen rather than in the form of a tent. The abdomen has dorsal cross bands on the rear segments and dark marginal plates.

Genus *Tinocallis*

The 22 species in this genus all live on Ulmaceae. Viviparous females are all winged and show great seasonal differences in pigmentation. They characteristically have both dorsal and marginal tubercles on the abdomen. The antennae can be either as long as the body or shorter. The cauda is knobbed, the siphunculi are stump-shaped with a broad base, and the anal plate is two-lobed. When disturbed the aphids escape by jumping, possibly using the muscles in the greatly enlarged coxae of the first pair of legs.

Tinocallis platani (Kaltenbach) **(Pl. 16.2, 16.3)**

This holocyclic, non-host-alternating species lives on elms (*Ulmus*). The aphids colonize the undersides of young leaves. Winged viviparous females are yellow or greenish white, with black or brown markings on the head, thorax, abdomen and forewings. The siphunculi are black. The head and pronotum have three dark stripes along them, and the tips of the antennal segments are dark. There are marginal plates and dark spots or cross bands on abdominal segments 3–6, and rather short finger-shaped spinal tubercles on abdominal segments 1 and 2. The antennae are about 0.75 times the length of the body, with a terminal process that is 0.2 times the length of the last antennal segment. The veins on the forewings have dark shadows and the radial sector is reduced.

Genus *Tuberculatus*

There are about 51 species in this genus and all live on oak. The viviparous females are always winged and characteristically have spinal tubercles on the abdomen. The antennae are as long as the body or longer. The cauda is knobbed, the siphunculi are short, truncate and smooth, and the anal plate is two-lobed.

Tuberculatus annulatus (Hartig) **(Pl. 16.4)**

This holocyclic, non-host-alternating species lives on oaks (especially *Quercus robur*). The aphids colonize the undersides of young leaves forming small groups or well spaced out, and are not attended by ants. Winged viviparous females are yellowish, greyish green, or pink to purple (in summer), with the tips of the antennal segments and of the spiphunculi dark. There are pairs of often dusky spinal tubercles or projections on abdominal segments 1–3. Those on segment 3 are particularly large. The antennae have a terminal process that is 0.7–1.1 times the length of the last antennal segment.

Tuberculatus borealis (Krzywiec)

This holocyclic, non-host-alternating species lives on oaks (especially *Quercus robur*). The aphids colonize the undersides of young leaves forming small colonies or well spaced out and are not attended by ants. They are similar morphologically to *T. annulatus*. Winged viviparous females are mottled green and yellowish, with the tips of the antennal segments and of the spiphunculi dark. There are pairs of often dusky spinal tubercles or projections on abdominal segments 1–4, with those on abdominal segments 1–3 all the same length and those on segment 4 sometimes very small. The antennae have a terminal process that is 0.9–1.3 times the length of the last antennal segment.

Tuberculatus neglectus Krzywiec

This holocyclic, non-host-alternating species lives on oaks (especially *Quercus petraea*). They colonize the undersides of young leaves. Winged viviparous females are yellowish, with a yellow mesothorax. The tips of the antennal segments and siphunculi are brown or black. The pairs of often dusky spinal tubercles or projections on abdominal segment 1–3 are all the same length. Abdominal segments 1–7 have wart-like marginal tubercles. The antennae are usually a little longer than the body and have a terminal process that is 1.0–1.6 times the length of the last antennal segment.

Genus *Tuberolachnus*

There are only two species in this genus and both characteristically have a large conical spinal tubercle or projection on the upper surface of abdominal segment 4.

Tuberolachnus salignus (Gmelin) **Large willow aphid (Pl. 16.5, 16.6)**

This anholocyclic, non-host-alternating species lives on the branches and twigs of willows (*Salix* species). Wingless viviparous females are dark brown, with black siphuncular cones. The body is covered with numerous fine hairs, which give it a greyish tinge. A large black tubercle is located on the back just in front of the siphunculi. The antennae are about 0.40–0.45 times the length of the body, with a terminal process that is 0.4 times the length of the last antennal segment. The forewing of winged individuals is unpigmented, with a slender, pointed pterostigma and a short nearly straight radial sector.

9 Techniques and approaches to original work

The natural environment is the ecological theatre in which aphids have evolved, and for a better understanding of their biology it is essential to observe and study them in nature. Below are a few tips for collecting and observing them in the field, and suggestions of how to go about experimenting with them in the field and laboratory. You will undoubtedly be fascinated and intrigued by what you see and record. After a time patterns will appear in the results and information you have amassed. For a deeper understanding it is important that you try to identify the processes that have shaped the patterns you identify and continually ask 'Why this particular pattern?' It is important to test your ideas experimentally. This is often very difficult in the field because it is not possible to ask a simple question of an aphid. So many things vary. Because of this most experiments are done in the laboratory or in semi-natural conditions. Wherever the experiments are done, it is important to keep an open mind and see the results as a way of stimulating further 'why' questions about the system you are studying.

Collecting and observing aphids in the field

The height of mature trees restricts one to studying the aphids living on the leaves and twigs of the lower branches. Although one should not assume that what happens in the lower canopy necessarily occurs in the upper canopy it nevertheless offers a convenient window for studying aphids living in the canopy of trees.

For collecting you need containers for transporting aphids and plant material back to the laboratory for closer examination or for experimental work. Small plastic or tin containers are the most suitable. Polythene bags can be used, but remember to enclose some air before sealing the bag. The entrapped air will prevent the contents from being crushed. It is important, especially in summer, to keep these containers in the shade; exposure to solar radiation will rapidly warm them up and reduce the contents to a soggy and unrecognizable mass.

The easiest way to collect aphids for identification is to transfer them directly into specimen tubes (about 5.5 x 1.5 cm) containing 70% industrial methylated spirits, using a very fine artist's paint brush (grade 00). After a little practice you will find it very easy. The spirit both kills and preserves the aphids.

A hand lens is essential. There are many types. Those that fold into a metal or plastic holder are the most convenient, especially those that consist of 2 or 3 separate lenses, which when used on their own or in combination give a range of magnification of between 10 and 20 times.

Above all, the most important feature for field use is a wide aperture (approximately 4 cm).

Rearing aphids in the laboratory

For laboratory studies you will need healthy plant material. Cutting leaf-bearing twigs from trees and placing the cut end in water will provide an ephemeral source of plant material. The best solution is to use young saplings. These can be grown from seed or by collecting germinating seedlings from below forest trees in spring. Alternatively, you can purchase 2- or 3-year-old saplings from a forest nursery. The saplings should be individually planted in pots, which can be kept outside, preferably with the pots sunk in soil and watered regularly during summer. In this way it is possible to maintain a supply of saplings of a range of sizes, which can easily be dug up and brought into the laboratory when required. Sycamore is particularly suitable for such studies, but grows very rapidly and quickly becomes a small tree.

To prevent the escape of winged aphids from saplings you can cover them with cages made of flexible clear PVC sheeting bent into a cylinder of the desired dimensions and glued with a suitable plastic adhesive (fig. 33). The open top can be sealed with a muslin lid glued around the edge. This keeps the aphids from escaping but allows some circulation of air, which prevents condensation. Alternatively, cages can be obtained from entomological suppliers (p. 127).

The growth, development and reproduction of individual aphids are most easily followed by caging them in clip cages (fig. 34). These are easily constructed from a spring-loaded ladies' metal hair clip, two short lengths of rigid Perspex tubing and two rings of foam rubber. First cut two rings from the Perspex tubing; they can be of the same depth or one can be shallower than the other. Glue fine muslin onto the outer side of each ring and a ring of foam rubber around the inner rim of each ring. The pointed ends of the prongs of the hairclips are heated gently in a flame until they are just glowing and then pushed through the wall of the Perspex ring and held firmly until they set into position. This takes a little skill, which is quickly achieved by trial and error. Alternatively, the prongs of the hairclip can be bent so that they fit neatly over and under the two valves of the clip cage, and then glued to the upper surface of the upper valve and the lower surface of the lower valve, respectively. The result is a very simple and easily-used cage, which can be used to confine aphids on individual leaves for

PVC cylinder

nylon mesh

Fig. 33. A simple insect cage covering a potted sapling.

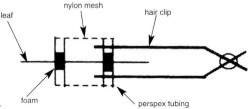

nylon mesh

leaf

hair clip

foam

perspex tubing

Fig. 34. A clip cage on a leaf.

long periods. In addition, this basic design can easily be modified so that the cage can accommodate more than one aphid, or aphids in both the upper and lower valves of the cage, or by the use of dividers the aphids can be kept physically separate from one another.

Experiments

One way to get a feel for what field and laboratory experiments can achieve is to read some easily available published studies on the sycamore aphid. Reading the relevant literature is an essential first step in any scientific investigation, and sycamore is a particularly suitable tree for such studies because its leaves are relatively large and the lamina of each leaf is normally horizontal.

Aggregation At any one time, more leaves on a sycamore tree lack aphids than one would expect on the basis of chance (Dixon, 1966). The aphids are therefore aggregated on certain favoured leaves. Why should this be the case? On the basis of a laboratory study, Kennedy & Crawley (1967) argued that the tendency to aggregate on certain leaves is self imposed and results in the regulation of numbers at a level well below the carrying capacity of their host trees. Although this idea was generally well received, its relevance clearly needed to be tested in the field. In Dixon & McKay (1970) you will find an account of one way of testing this idea in the field. This study reveals that leaves differ greatly in terms of the associated microenvironment and the aphids tend to settle on those leaves, or parts of leaves, that have a favourable microenvironment, and avoid those with an unfavourable microenvironment (fig. 35).

Spacing A sycamore aphid kicks when closely approached by another aphid, which then often moves away. This results in the spacing out of the aphids. Although they repel one another, the aphids also attract one another and settle so that they can just touch their neighbours with their antennae. Kennedy & Crawley (1967) describe this as 'spaced-out gregariousness', and they conclude that sycamore aphids

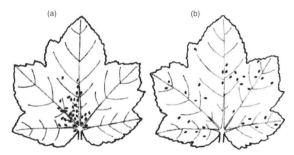

Fig. 35. The distribution of aphids on a sycamore leaf in the field (a) during a high wind and (b) after the high wind abated.

space themselves so that there is a definite distance, independent of density, at which each aphid prefers to maintain its nearest neighbour. This concept was challenged by Dixon & Logan (1972), who showed that when there are many aphids on a leaf they are more closely packed together than when there are few, and that vision does not appear to play a part in the spacing.

Availability of space As a rule the sycamore aphid settles on the under surface of a leaf and on visible veins, which are the source of its food. The veins of a leaf vary in size as do the aphids, which are smallest at birth and largest when adult. Just how the relative sizes of aphids and veins determine the distribution and availability of space for the sycamore aphid is explored by Dixon & Logan (1973).

Similarly, it is possible to determine the factors that prevent aphids settling and feeding on the upper surface of leaves and on petioles during summer. Dixon (1976) showed that the distribution of sycamore aphids reflects their response to seasonal changes in the quality of the food available when feeding on different parts of a leaf, and the microenvironment at the leaf surface.

These examples show that close observation of the distribution of aphids on the leaves of trees over a period of a few weeks, followed by simple manipulation of the shape and position of the leaves, can greatly increase our understanding of why aphids prefer to occupy certain leaves and parts of leaves, and can help us to appreciate the complexity of the environment aphids live in. All that is required is curiosity and a keen desire to answer the question 'why?'.

Checklist of aphid genera

A checklist of the aphid genera keyed out in this book, according to the classification of Remaudiere & Remaudiere (1997). All belong to the family Aphidae.

Subfamily	Tribe	Subtribe	Genus
Aphidinae	Aphidini	Aphidina	*Aphis*
		Rhopalosiphina	*Rhopalosiphum*
	Macrosiphini		*Anuraphis*
			Cavariella
			Corylobium
			Dysaphis
			Myzus
			Ovatus
Calaphidinae	Calaphidini		*Betulaphis*
			Calaphis
			Callipterinella
			Clethrobius
			Crypturaphis
			Euceraphis
			Monaphis
			Symydobius
	Panaphidini		*Chromaphis*
			Eucallipterus
			Myzocallis
			Panaphis
			Pterocallis
			Tinocallis
			Tuberculatus
Chaitophorinae	Chaitophorini		*Chaitophorus*
			Periphyllus
Drepanosiphinae			*Drepanosiphum*
Eriosomatinae	Eriosomatini		*Colopha*
			Eriosoma
			Kaltenbachiella
			Tetraneura
Lachninae	Lachnini		*Lachnus*
			Stomaphis
			Tuberolachnus
Pemphiginae	Pemphigini		*Pachypappa*
			Patchiella
			Pemphigus
			Prociphilus
			Thecabius
Phyllaphidinae			*Phyllaphis*
Pterocommatinae			*Plocamaphis*
			Pterocomma
Thelaxinae			*Thelaxes*

Some useful addresses

Societies and publications

The Amateur Entomologists' Society, P.O. Box 8774,
London SW7 5ZG
www.theaes.org

The British Entomological and Natural History Society, The
Pelham-Clinton Building, Dinton Pastures Country Park,
Davis Street, Hurst, Reading, Berkshire RG10 0TH
www.benhs.org.uk

The Royal Entomological Society, The Mansion House,
Chiswell Green Lane, Chiswell Green, St Albans AL2 3NS
(tel 01727 899387) www.royensoc.demon.co.uk

For AIDGAP keys, Royal Entomological Society Handbooks
for the Identification of British Insects, and Linnean Society
Synopses of the British Fauna: Field Studies Council, FSC
Publications, Preston Montford, Montford Bridge,
Shrewsbury, Shropshire SY4 1HW (tel 0845 3454072)
www.field-studies-council.org

For Naturalists' Handbooks: The Company of Biologists Ltd,
140 Cowley Road, Cambridge CB4 0DL
(tel 01223 426164/420482)
sales@biologists.com, www.biologists.org

Richmond Publishing Company Ltd: P.O. Box 963,
Slough SL2 3RS (tel 01753 643104)
e-mail rpc@richmond.co.uk

Equipment suppliers

For field and laboratory equipment: Philip Harris Education,
Customer Service Centre, Novara House, Excelsior Road,
Ashby Park, Ashby De La Zouch, Leicestershire LE65 1NG
(tel 0870 6000193) www.philipharris.co.uk

For equipment for catching, setting and preserving insects:
Watkins and Doncaster, P.O. Box 5, Cranbrook, Kent
TN18 5EZ (tel. 01580 753133) www.watdon.com

For materials for microscopy, slides, insect pins etc.:
D. J. and D. Henshaw, 34 Rounton Road, Waltham Abbey,
Essex EN9 3AR (tel/fax 01992 717663)
e-mail djhagro@aol.com

For rigid plastic tubing: Abbey Distribution Ltd, Abbey
House, Lisle Road, High Wycombe HP13 5SH
(tel 01494 449975) www.abbeydistribution.com

For carbon dioxide dispensers: Edme Ltd, Customer Services
Department, Edme Ltd, Mistley, Manningtree, Essex
CO11 1HG (tel 01206 393725)

References and further reading

Akimoto, S. (1988) Competition and niche relationships among *Eriosoma* aphids occurring on Japanese elm. *Oecologia* **75**: 44–53.

Antolin, M.K. & Addicott, J.F. (1988) Habitat selection and colony survival of *Macrosiphum valeriani* Clarke (Homoptera: Aphididae). *Annals of the Entomological Society of America* **81**: 245–251.

Blackman, R. (1974) *Aphids*. London: Ginn & Company Ltd.

Blackman, R.L. & Eastop, V.F. (1994) *Aphids on the world's trees. An identification and information guide*. Wallingford: CAB International.

Blackman, R.L. & Eastop, V.F. (2000, 2nd edn.) *Aphids on the world's crops. An identification and information guide*. Chichester: John Wiley & Sons.

Bonnet, C. (1745) *Traité d'insectologie ou observations sur les pucerons*. Paris: Chez Durand

Börner, C. (1939) Anfälligkeit, Resistenz und Immunität der Reben gegen Reblaus. Allgemeine Gesichtspunkte zur Frage der Spezialisierung von Parasiten: die harmonische Beschränkung des Lebensraums. *Zeitschrift für Hygiene Zoologie Schädlingsbekampfung* **31**: 274–285, 301–308, 325–334.

Bumroongsook, S. & Harris, M.K. (1992) Distribution, conditioning and interspecific effects of black-margined aphids and yellow pecan aphids (Homopetra: Aphididae) on pecan. *Journal of Economic Entomology* **85**: 187–191.

Büsgen, M. (1891) Der Honigtau. Biologische Studie an Pflanzen und Pflanzenläusen. *Jenaische Zeitschrift für Naturwissenschaft* **25**: 339–428.

Calunga-Garcia, M. & Gage, S.H. (1999) Arrival, establishment, and habitat use of the multicolored Asian ladybird beetle (Coleoptera: Coccinellidae) in a Michigan landscape. *Environmental Entomology* **27**: 1574–1580.

Dixon, A. F. G. (1958) The protective function of the siphunculi of the nettle aphid *Microlophium evansi* (Theob.) (Hem. Aphididae). *Entomologist's Monthly Magazine* **94**: 8.

Dixon, A.F.G. (1966) The effect of population density and nutritive status of the host on the summer reproductive activity of the sycamore aphid, *Drepanosiphum platanoides* (Schr.). *Journal of Animal Ecology* **35**: 105–112.

Dixon, A.F.G. (1976) Factors determining the distribution of sycamore aphids on sycamore leaves during summer. *Ecological Entomology* **1**: 275–278.

Dixon, A.F.G. (1998, 2nd edn.) *Aphid ecology*. London: Chapman & Hall.

Dixon, A.F.G. (2000) *Insect predator–prey dynamics: ladybird beetles and biological control*. Cambridge: Cambridge University Press.

Dixon, A.F.G. (2005) *Insect herbivore–host dynamics: tree-dwelling aphids*. Cambridge: Cambridge University Press.

Dixon, A.F.G., Kindlmann, P., Leps, J. & Holman, J. (1987) Why are there so few species of aphids, especially in the tropics? *American Naturalist* **129**: 580–592.

Dixon, A.F.G. & Logan, M. (1972) Population density and spacing in the sycamore aphid, *Drepanosiphum platanoides* (Schr.), and its relevance to the regulation of population growth. *Journal of Animal Ecology* **41**: 751–759.

Dixon, A.F.G. & Logan, M. (1973) Leaf size and availability of space to the sycamore aphid *Drepanosiphum platanoides*. *Oikos* **24**: 58–63.

Dixon, A.F.G. & McKay, S. (1970) Aggregation in the sycamore aphid *Drepanosiphum platanoides* (Schr.) (Hemiptera: Aphididae) and its relevance to the regulation of population growth. *Journal of Animal Ecology* **39**: 439–454.

Eastop, V. (1972) Deductions from the present day host plants of aphids and related insects. In *Insect/Plant Relationships: Symposia of the Royal Entomological Society of London* **6:** 157–178.

Edson, J.I. (1985) The influence of predation and resource subdivision on the coexistence of goldenrod aphids. *Ecology* **66**: 1736–1743.

Ehrlich, P.R. & Raven, P.H. (1964) Butterflies and plants: a study in coevolution. *Evolution* **18:** 586–608.

Elton, C.S. (1925) The dispersal of insects to Spitsbergen. *Transactions of the Entomological Society of London* **1925:** 289–299.

Evans, E.W. (2004) Habitat displacement of North American ladybirds by an introduced species. *Ecology* **85:** 637–647.

Fisher, R.A. (1930) *The genetical theory of natural selection*. New York: Dover.

Foster, W.A. & Northcott, P.A. (1994) Galls and the evolution of social behaviour in aphids. In *Plant galls* (ed. M.A.J. Williams), pp. 161–182. Oxford: Clarendon Press.

Furuta, K. (1986) Host preference and population dynamics in an autumnal population of the maple aphid *Periphyllus californiensis* Shinji (Homoptera, Aphididae). *Zeitschrift für angewandte Entomologie* **102**: 93–100.

Furuta, K. (1990) Early budding of *Acer palmatum* caused by the shade; intra-specific heterogeneity of the host for the maple aphid. *Bulletin of the Tokyo University Forests* **82**: 137–145.

Gilbert, F.S. (1993) *Hoverflies*. Naturalists' Handbooks 5. Slough: The Richmond Publishing Co. Ltd.

Hajek, A.F. & Dahlsten, D.L. (1986) Coexistence of three species of leaf-feeding aphids (Homoptera) on *Betula pendula*. *Oecologia* **68**: 380–386.

Hamilton, W.D. (1967) Extraordinary sex ratios. *Science* **156**: 477–488.

Heie, O.E.: The Aphidoidea (Hemiptera) of Fennoscandia and Denmark:
(1980) I. General part. The families Mindaridae, Hormaphididae, Thelaxidae, Anoeciidae, and Pemphigidae. *Fauna entomologica scandinavica* **9**: 236 pp.

(1982) II. The family Drepanosiphidae. *Fauna entomologica scandinavica* **11**: 176 pp.

(1986) III. The family Aphididae: subfamily Pterocommatinae & tribe Aphidini of subfamily Aphidinae. *Fauna entomologica scandinavica* **17**: 314 pp.

(1991) IV. The family Aphididae: Part 1 of tribe Macrosiphini of subfamily Aphidinae. *Fauna entomologica scandinavica* **25**: 188 pp.

(1993) V. The family Aphididae: Part 2 of tribe Macrosiphini of subfamily Aphidinae. *Fauna entomologica scandinavica* **28**: 239 pp.

(1995) VI. The family Aphididae: Part 3 of tribe Macrosiphini of subfamily Aphidinae. *Fauna entomologica scandinavica* **31**: 217 pp.

Huxley, T.H. (1858) On the agamic reproduction and morphology of *Aphis*. Part 1. *Transactions of the Linnean Society* **22:** 193–219.

Inbar, M. & Wool, D. (1995) Phloem feeding specialists sharing a host tree: resource partitioning minimizes interference competition among galling aphid species. *Oikos* **73**: 109–119.

Jackson, D.L. & Dixon, A.F.G. (1996) Factors determining the distribution of the green spruce aphid, *Elatobium abietinum*, on young and mature needles of spruce. *Ecological Entomology* **21**: 358–364.

Kennedy J.S. & Crawley, L. (1967) Spaced-out gregariousness in sycamore aphids *Drepanosiphum platanoides* (Schr.) (Hemiptera: Callaphididae). *Journal of Animal Ecology* **36**: 147–170.

Knäbe, S. & Dixon, A.F.G. (2004) Symbionts and host specificity in aphids. In *Aphids in a new millennium*, (eds. J.C. Simon, C.A. Dedryver, C. Rispe & M. Hullé), pp. 457–462. Paris: INRA.

Kurosu, U., Aoki, S. & Fukatsu, T. (2003) Self-sacrificing gall repair by aphid nymphs. *Proceedings of the Royal Society of London* B **270**: (Suppl.1) S12–S14.

Lees, A.D. (1960) The role of photoperiod and temperature in the determination of parthenogenetic and sexual forms in the aphid *Megoura viciae* Buckton. II. The operation of the "interval timer" in young clones. *Journal of Insect Physiology* **4**: 154–175.

Majerus, M. & Kearns, P. (1989) *Ladybirds*. Naturalists' Handbooks 10. Slough: The Richmond Publishing Co. Ltd.

Nieto Nafria, J.M., Mier Durante, M.P. & Remaudiere, G. (1997) Les noms des taxa du group-famille chez les Aphididae (Hemiptera). *Revue francaise d'entomologie (N.S.)*, **19** (3–4): 77–92.

Rakauskas, R. (2004) Recent changes in aphid (Hemiptera, Sternorrhyncha: Aphididae) fauna of Lithuania: an effect of global warming? *Ekologija* **1**: 1–4.

Réaumur, R.A.F. de (1937) *Mémoires pour servir a l'histoire des insectes* **2 Vols,** Paris: De L'Imprimiere Royale.

Remaudiere, G. & Remaudiere, M. (1997) *Catalogue des Aphididae du Monde. Catalogue of the world's Aphididae*. Paris: INRA, 473 pp.

Rotheray, G.E. (1989) *Aphid predators*. Naturalists' Handbooks 11. Slough: The Richmond Publishing Co. Ltd.

Scarborough, C.L., Ferrari, J. & Godfray, H.C.J. (2005) Aphid protected from pathogen by endosymbiont. *Science* **310**: 1781.

Shaposhnikov, G. Kh. (1964) Aphidinea in: Keys for the Identification of the Insects of the European part of the USSR (ed. Bei-Bienko), **1**: 489–616.

Stadler, B. & Dixon, A.F.G. (2005) Ecology and evolution of aphid–ant interactions. *Annual Review of Ecology, Evolution and Systematics* **36**: 345–372.

Stern, D.L. & Foster, W.A. (1996) The evolution of soldiers in aphids. *Biological Reviews* **71**: 27–79.

Strong, D.R., Lawton, J.H. & Southwood, T.R.E. (1984) *Insects on plants*. Oxford: Blackwells.

Stroyan, H.L.G. (1977) Homoptera, Aphidoidea: Chaitophoridae and Callaphididae. *Handbooks for the identification of British insects* **2** (Part 4a). London: Royal Entomological Society of London.

Stroyan, H.L.G. (1984) Aphids: Pterocommatinae and Aphididinae (Aphidini) Homoptera, Aphididae. *Handbooks for the identification of British insects*, **2** (Part 6). London: Royal Entomological Society of London.

Tamas, I., Klasson, L., Canback, B., Näslund, A.K., Eriksson, A.-S., Wernegreen, J.J., Sandström, J.P., Moran, N.A. & Andersson, S.G.E. (2002) 50 million years of genomic stasis in endosymbiotic bacteria. *Science* **296**: 2376–2379.

Thieme, T. & Müller, F.P. (2000) Unterordnung Aphidina: Blattläuse, Aphiden. In *Exkursionsfauna von Deutschland*, Vol. **2**, *Wirbellose: Insekten* (eds. H.-J. Hannemann, B. Klausnitzer & K. Senglaub). Jena: Spektrum Akademische Verlag.

Vidano, C. (1959) Analisi morfologica ed etologica del ciclo eterogonico de *Rhopalosiphum oxyacanthae* (Schrank) Börner su Pomoidee e Graminacee (Hemiptera Aphididae). *Bolletino de Zoologia Agraria e Bachicoltura*, **2** (2): 1–225.

Völkl, W. (1989) Resource partitioning in a guild of aphid species associated with creeping thistle, *Cirsium arvense*. *Entomologia Experimentalis et Applicata* **51**: 41–47.

White, G. (1887) *The natural history of Selborne*. London: Walter Scott.

132

INDEX